T0326393

Unlocking markets to smallholders

Unlocking markets to smallholders

Lessons from South Africa

edited by:
Herman D. van Schalkwyk
Jan A. Groenewald
Gavin C.G. Fraser
Ajuruchukwu Obi
Aad van Tilburg

Mansholt publication series - Volume 10

ISBN: 978-90-8686-134-7
e-ISBN: 978-90-8686-168-2
DOI: 10.3920/978-90-8686-168-2

ISSN: 1871-9309

Cover idea:
Ajuruchukwu Obi

Cover photography:
Upper-left: SA farmers near Lesotho going to the market (photo Andries Jordaan)
Bottom-left: cows (to be) fattened for the market (Free State) (photo Aad van Tilburg)
Upper-central photo: lambs on bottle feeding (farm near Clarence) (photo Aad van Tilburg)
Bottom-central: Ploughing farmers near Lesotho (photo Andries Jordaan)
Right-side: Emerging farm in the Free State (photo Aad van Tilburg)

First published, 2012

Mansholt Publication Series

The Mansholt Publication Series (MPS) contains peer-reviewed publications on social changes, transformations and control processes in rural areas and (agri)food chains as well as on their institutional contexts. MPS provides a platform for researchers and educators who wish to increase the quality, status and international exposure of their scholarly work.

The Series is named after Sicco Mansholt (1908-1995), who was Minister of Agriculture in The Netherlands from 1945 until 1958. In addition he was the European's Commissioner of Agriculture and Vice-President of the European Commission from 1958 until 1972.

MPS is supported by the Wageningen School of Social Sciences (WASS), the merger of former Mansholt Graduate School of Social Sciences and CERES Research School for Resource Studies for Development. The quality and contents of the Series is monitored by an interdisciplinary editorial board. All submitted manuscripts are reviewed by at least two independent reviewers before being considered for publication. MPS is published and marketed internationally by Wageningen Academic Publishers.

The Mansholt Publication Series editors are:

Preface

This book documents the key findings of a study designed to investigate the institutional and technical constraints to smallholder agriculture in selected areas of South Africa. To what extent can small-scale farmers share in the expected gains of integration into national or international markets and what institutional and other reforms are necessary to enhance their effective and profitable participation in the regional economy?

In order to answer this question, the book attempts to shed some light into the diverse factors that affect access to output markets for South Africa's small-scale farmers or their access to input markets where they procure needed inputs and farm services (e.g. credit). South African smallholders are predominantly black farmers who generally started their business from a situation in which they lacked basic resources such as economic, social and human capital. The ideal is to attain a situation in which these smallholders become more productive, more market-oriented and better connected to markets than before. Some categories of these smallholders, usually called emerging farmers, are more market oriented than others because they produce purposely for local, national or international markets. With improved knowledge about the opportunities and constraints facing smallholders or emerging farmers, private and public policy makers at national and provincial levels will be better equipped to focus their support activities on the needs of these smallholders or emerging farmers with the aim to improve their relative positions in the value chain.

The original ideas for the study – as represented in this book – were hatched at the Dean's office of the Faculty of Natural and Agricultural Sciences of the University of the Free State, Bloemfontein, South Africa in 2003 where Herman van Schalkwyk and Ajuruchukwu Obi developed a concept note for a research on 'Assessing the institutional and technical constraints to smallholder agriculture in selected areas of South Africa and implications for market access, poverty alleviation, and socio-economic sustainability'. This note was shared with Aad van Tilburg who made inputs leading up to the elaboration of a project proposal that was submitted to the South Africa Netherlands Research Programme on Alternatives in Development (SANPAD) and was approved in 2004 as SANPAD-project 04/11. Gavin Fraser was invited to participate in a start-up workshop during which a project team of the four persons mentioned was formally established to initiate and supervise the study. Jan Groenewald joined the team when the ideas about a book were put into practice. The book consists of contributions by the five editors and both colleagues and (former) PhD students of the University of the Free State as well as former MSc students of both the University of the Free State and the University of Fort Hare.

The editors and authors of the book express their gratitude to the editors of the Mansholt Series of Wageningen Academic publishers for the inclusion of this book in their series and to the peer reviewer who gave highly valuable comments and suggestions for improvement of the manuscript in two rounds.

The book has been written with the following audience in mind:
- members of the farming and institutional community in South Africa being or supporting smallholders or emerging farmers;
- the South African agribusiness community that is – with the support of Black Economic Empowerment regulation – integrating the smallholder sector in their activities;
- South African policy makers with respect to smallholder farming or emerging farmers;
- students and scholars in the domain of agriculture, agricultural economics and regional development in South Africa; and
- similar categories outside South Africa with a keen interest on how to improve market access of smallholder farmers.

The editors
South Africa/the Netherlands
September 2011

Table of contents

1. Market access, poverty alleviation and socio-economic sustainability in South Africa

Ajuruchukwu Obi, Herman D. van Schalkwyk and Aad van Tilburg

1.1 Introduction

Markets continue to be seen as the means for ensuring that smallholder producers of agricultural products are effectively integrated into the mainstream of national economies, especially in developing countries. For one thing, markets provide the opportunity for farm production to contribute to poverty reduction through the cash income realised from sales of farm produce. In turn, markets drive production as farmers strive to meet the demands of consumers and end-users in terms of quantity and quality. But their very existence, or how effectively they function, cannot be guaranteed in many developing countries. In South Africa, there is a certain urgency to address the real concern that, in spite of considerable investments into restructuring the sector since 1994 and directly tackle agrarian and land reform, poverty is still rife and there is the clear indication that much of this arises from farmers not being able to sell produce at a profit. Unlocking markets for this group of farmers is therefore considered a crucial developmental necessity. Research conducted from 2004 in various parts of the country point to the importance of the market access to smallholders. The aim of this book is to attempt an aggregation of the findings from an investigation into the technical and institutional constraints to smallholders' market access and how these affect other aspects of community life. Without a doubt, such concerns are not new and have formed part of theoretical and policy work focusing on the gains from trade for several centuries.

From the 16th century when the Mercantilists began to debate on the gains from trade, the issue of market access has been of more than passing interest to development theorists and policy makers alike. Apart from recognising the importance of trade in reducing differences in welfare between nations by building up their stock of economic assets, those early economic theorists conceptualised a role for trade also in creating and sustaining differences in economic well-being of nations when one country was able to expand its exports to another while simultaneously choking off imports from such a country. Introducing the notion of access to markets as well as of comparative advantage at the beginning of the 19th century no doubt arose from such a mindset. According to David Ricardo (1817), trade 'powerfully contribute(s) to increase the mass of commodities and therefore the sum of enjoyments'. Modern-day development planning, advisory and technical assistance have all tried to build on such knowledge to enhance the capacity of countries to bring about their own development by expanding the production of goods for which a vibrant market exists. In much the same way, regional and local development have been found to owe a great deal to the differences in production possibilities as well as the ease with which surplus produce can be marketed at a profit. The expectation, following Ricardo's thesis and its numerous modern versions, including the Heckscher and Ohlin Model and more recent trade theory of Paul Krugman (Krugman, 2008; Leamer, 1995), is that the discovery of new markets increases the value of a nation's goods and services thus add to the growth of national wealth.

While the foregoing views about markets remain unchanged, the current global environment is defining new roles for markets especially in developing and emerging economies. Globalisation has changed the dynamics of the agro-food systems to the extent that new questions are being asked about the role of the smallholder, what happens to traditional agriculture, how traditional markets are evolving, how roles are allocated between the private and public sectors, and the power relations among all the stakeholders in the global market arena (Gabre-Madhin, 2006). Despite the unquestionable importance of markets, the irony is that their existence is not always guaranteed; asymmetries in access to remunerative markets remain a crucial obstacle. The reality of present-day economic life in sub-Saharan Africa is that either markets fail to function efficiently, representing various degrees of market failures, or they simply disappear – the so-called missing markets.

The structural adjustment programmes (SAP) that were launched in many countries of the continent from the mid-1980s onwards were predicated on the notion that the fight against poverty is meaningless if farmers cannot raise revenues by selling what they produce at a profit. South Africa's return to the international arena occurred at a time when it was unavoidable to embrace that mindset, including its fundamental misjudgment that 'getting prices right' was all that was required to reverse market failures and bring markets into existence. According to Makhura and Mokoena (2003), several 'market deregulation' and 'trade liberalisation' policies were introduced by the government in the post-1994 era as part of the menu of strategies to link small farmers to agricultural markets. But these policies have met with limited success and the problem of insufficient market access for smallholders remains a serious concern today. Why is this so?

In the past decades, numerous studies have attempted to answer this question. Many of these studies have been designed to come up eventually with a listing of the key obstacles that confront smallholders who are unable to enter new markets or perform satisfactorily in existing ones. Organisations like the World Bank, the International Fund for Agricultural Development (IFAD) and the International Food Policy Research Institute (IFPRI) have been in the forefront of the efforts to understand the nature and extent of these problems. Bienabe *et al.* (2004), IFAD (2003), Minot and Hill (2007), and the World Bank (2007) have all concluded that four categories of constraints are important and include:

- constraints associated with high costs of transaction that raise marketing cost;
- constraints associated with the riskiness of agricultural production especially in respect to new products;
- constraints associated with the riskiness of marketing of agricultural produce in the presence of poor infrastructure and/or high price variability;

- constraints associated with weakening of primary markets and/or bargaining power of producers and sellers.

In the context of developing countries and the current global concern with poverty alleviation, hardly any sector has received as much attention in relation to market access as the agricultural sector. The reasons for this are obvious. In the first place, agriculture remains a dominant sector for these countries in terms of its share of national wealth and employment. This means that any actions to enhance the livelihoods of the population must either be based on agriculture or have strong links with that sector. Secondly, as a result of limited opportunities elsewhere in these economies, efforts to expand employment opportunities, at least in the short-term, must be pivoted on agriculture. How the sector is faring in terms of gross production, what constraints are encountered in the production process, and the opportunities for marketing, thus assume crucial practical implications for national policy to achieve sustained economic growth and poverty alleviation in all sectors of the economy including the communal areas.

The purpose of this chapter is to set the context for such an enquiry into the performance of the agricultural sector of South Africa against the backdrop of its recent economic and political history. For a country, like South Africa that emerged from economic and political isolation since 1994 with the promise to pursue an all-inclusive development strategy at home while assuming regional and continental leadership aiming at greater economic integration, such an analysis cannot be avoided. The question as to why nearly two decades of reforms have not produced the dramatic change in rural livelihoods as expected has not yet been confronted systematically. The answers to these questions are even more urgent at this time when local discontent is brewing and the population is showing signs of impatience with what seems to be the slow pace of the reform process. In this regard, the spate of 'service-delivery protests' and industrial actions that have rocked the country since 2006, are stark reminders of the enormous burden that the democratic regime in South Africa must shoulder in order to achieve meaningful socioeconomic transformation.

Without question, the challenge of reform for South Africa is complicated by factors and developments that are exogenous to the system and are inevitable fall-outs of the greater openness that liberalisation and democratisation brought about. One of the most significant developments from the standpoint of poverty alleviation and socioeconomic sustainability is the global food crisis that came to a head in 2008. An unprecedented 50 odd years when world food supplies expanded astronomically on the wings of technological developments (triggered by the Green Revolution of the early 1960s), resulted in significant declines in real food prices (Southgate 2010; Southgate *et al.* 2007). Ironically, these gains have been largely reversed by economic progress since March 2007. Several analysts have suggested that the phenomenal growth in energy demand, especially in emerging markets such as China and India, has fuelled increases in oil prices, which in turn raised the costs of vital farm inputs such as fertilisers, diesel, and other inputs (FAO, 2008; Southgate, 2010). The last major commodity price increase in 1973-1974 followed the Middle East oil shock of the early 1970s (Church, 2005; Green, 1978; Wakeford, 2006), and subsequently oil shocks have led to similar food price increases. Of course, there is the now famous demand side culprit of biofuel development, particularly the production of ethanol from maize, which is causing food prices to spike (Masters, 2008; Southgate, 2010). Estimates summarised by USAID (2009) show that overall increase in food prices averaged 43% in the one year period beginning in March 2007 up to the first quarter of 2008, being most serious for wheat, soybean, maize and rice which rank among the most important staples in the developing countries.

According to Aerni (2008) and Masters (2008), among others, food price increases worsen poverty even if some poor people also benefit from rising food prices. Data generated by USAID show that as many as 8.7 million people in Southern Africa were most severely affected by these food price increases to the point of starvation or related hardships. While most of these persons were in the poorer countries of the region, such as Zimbabwe, Lesotho and Swaziland, there are many parts of South Africa, including the Eastern Cape, where real hunger stares people in the face, forcing the government to adopt food aid schemes in the form of distribution of food parcels and social grants. Whereas in 2003, 6.8 million South Africans were recipients of social grants, by 2010 the figure has grown to 14 million (Department of Social Development, 2010). Further evidence of the negative impact of the high food prices is that government has increased the social relief budget, that is the support provided to persons that fall into destitution as a result of food price increases, from R124 million to R624 million between 2008 and 2009 (Department of Social Development, 2010). There is also evidence that child hunger, diseases and deaths are taking place in many parts of the country as a direct consequence of high food prices as unemployed or low-wage persons are finding it increasingly difficult to afford basic necessities such as milk, meat and even cereals.

This chapter initially examines the background and problem context and then more comprehensively reviews the relevant literature on market access, poverty alleviation, and socio-economic sustainability. These issues are situated within the country's dualistic structure and the plethora of policy initiatives currently underway to redress mistakes of the past. Since the book is the product of a systematic investigation and research process, the chapter also provides insights into the research methodology employed and restates the fundamental argument that informed the research effort in the first place. The concluding part of the chapter is devoted to a description of the three provinces of South Africa in which the bulk of the research was undertaken in order to show the prior conditions that constituted the justification for the study in the first instance. This is then followed by chapter summaries that highlight the main findings of the studies.

1.2 Problem context

In the nearly two decades since democratic governance was enthroned in South Africa, the circumstances of the smallholder agricultural sector of the country have attracted considerable attention. With the knowledge that policies promoted before 1994 confined this smallholder sector to a mere 13% of the available agricultural land, there was already considerable urgency on the part of the new policy élite to right past wrongs. The parts of the country previously designated 'independent homelands' and 'self-governing territories' formed the living and farming environments of the majority black population and featured extreme cases of infrastructure deficiencies. According to Van Zyl and Binswanger (1996), these former homelands were characterised by inadequate market access, infrastructure and support services. Data available at the inception of democratic rule suggest that three million black farmers lived in these former homelands and produced food mostly for subsistence on much less than 17 million hectares of largely unfertile land (National Department of Agriculture, 2001). This contrasted with the about 50,000 large-scale commercial farmers who owned or operated about 102 million hectares of land (Ortmann and Machethe, 2003; Van Schalkwyk, 1995). A study conducted in 1999 had put the number of commercial farmers at about 60,000 and suggested that these farmers controlled 86% of the country's land while an estimated 1.25 million black people operated at subsistence levels on about 14% of the available land (Eicher and Rukuni, 1999). The importance of commercial farming in South Africa is underlined by the fact that they generated an export value of about US$ 2.3 billion in 2000 (Ortmann and Machethe, 2003). Although much of the pre-1994 government support that almost exclusively targeted the large-scale commercial farms has been discontinued, some even before political liberalisation, their effects are still well-entrenched and very much around in the form of well-laid infrastructural networks and know-how built over many years with substantial state subsidisation (Mbongwa *et al.*, 1996; Van Schalkwyk *et al.*, 2003).

The foregoing scenario defines the dualism of South African's agriculture and presents serious policy-making challenges. The government has embarked on a comprehensive land reform programme aiming to narrow the yawning gap in land ownership (Makhura and Mokoena, 2003; Van Schalkwyk, 1995; Van Zyl and Binswanger, 1996). Obviously, because of the high cultural significance of land and the deep emotions surrounding land matters (Obi, 2006; Van Schalkwyk, 1995), the land reform programme has received a significant, if not disproportionate, amount of both official and media attention. But land ownership, though a serious matter and obviously politically-sensitive and highly emotive given the history, is only one dimension of the problems facing the previously-disadvantaged communities of South Africa. According to Lyne (1996), smallholder farmers are further constrained by limited access to the other factors of production, including credit facilities and information. As Ortmann and Machethe (2003) noted, even with small farm sizes, agricultural productivity can be improved if adequate access to support services and markets is guaranteed.

Where poverty reduction is a central goal of economic policy, market access for producers assumes immense significance. How the food marketing system functions also has implications for the pace and level of regional development (Van Tilburg, 2003). So, the food marketing systems of developing countries have naturally been the subject of considerable academic and policy interest in recent times. Recent studies on the role of trade and market access have shown that significant gains can accrue to farmers if systems and procedures for the marketing of surplus produce are improved, especially in the African context (Rodrik, 1998; Roe, 2003; Van Schalkwyk and Jooste, 2003). The Government of South Africa has accordingly initiated far-reaching trade liberalisation and market deregulation processes alongside the other measures aiming to 'democratise' the agricultural sector (Makhura and Mokoena, 2003; Van Schalkwyk *et al.*, 2003).

However, more than 10 years into these reform measures, there has as yet been too little visible change in the circumstances of the rural small-scale producers of South Africa. The measures aimed at liberalising the domestic food market and integrating the country into the international system may actually have hurt rather than helped the small-scale farmers within the former homelands of South Africa (Makhura and Mokoena, 2003; Mokoena, 2002; Ndirangu, 2002; Nyamande-Pitso, 2001; Van Schalkwyk *et al.*, 2003). While the role of South African agriculture has grown in regional and international trade, it appears that small-scale farmers have hardly been part of this process (Makhura and Mokoena, 2003; Van Schalkwyk *et al.*, 2003). In many situations where surplus production has been realised by the rural small farmers, lack of access to markets has forced them into exploitative exchange arrangements, which further erode their welfare and drive them deeper into destitution.

Expectedly, the level of dissatisfaction among the populace is growing as the larger population confronts a reality that contrasts sharply with the lofty expectations they had in 1994. The Government is aware of this state-of-affairs. For example, at the presentation of the 2004 Budget Speech, the Finance Minister, Mr. Trevor Manuel, was bemoaning the emergence of a '...second economy characterised by poverty, inadequate shelter, uncertain incomes and the despair of joblessness...' in which many South Africans are currently 'trapped' (Government of South Africa, 2004). In desperation, Government is committing a sizeable chunk of the budget to the provision and upgrading of infrastructures serving local communities under a newly-created Municipal Infrastructure Grant (Department of Provincial and Local Government, 2003; Government of South Africa, 2004). There is an urgent need for empirical data and comprehensive information on the actual circumstances of the smallholders and the constraints they face to enable policy makers in programming future support initiatives, including the Municipal Infrastructure Grant, aiming to enhance the living conditions of the smallholders by raising their productivities and improving their market access.

For the purpose of the investigation into the constraints to smallholder market access in South Africa, a review of the relevant literature was considered necessary to establish the conceptual and theoretical basis for the current concern and subsequent actions that will need to be taken to correct the present situation. The literature review directly focuses on the importance of trade for agricultural and economic development, the role of agriculture in African economic recovery, the impact of marketing and agrarian reforms on smallholder agriculture, and the role of institutions and other public policy measures in enhancing market access and improving rural livelihoods. The next few sub-sections attempt to draw attention to the major highlights of the review.

1.2.1 Importance of agricultural trade and market access

At a broader level, the primacy of trade in the quest for economic development is no longer in dispute, and with globalisation its importance becomes even more pervasive (Rodrik, 1998; Summers, 1992; UNDP, 2003; Van Schalkwyk and Jooste, 2003; World Bank, 2002, 2003). Despite its challenges, globalisation does present immense opportunities for better development outcomes that can best be accessed through trade. For African countries whose principal economic activities are in agriculture, adding value by processing and trade is the only means for realising the benefits of globalisation (Van Schalkwyk, 2003). Even if the significant paradigm shifts of the late 1980s and early 1990's in the Eastern bloc countries are ignored, and regardless of the anti-market rhetoric of the professional protesters, there is hardly any question about the strong intellectual and, even ideological, support for, and merit of, market-led development strategies (Summers and Thomas, 1993). It is equally recognised that the 'gains from trade are a key source of national wealth' (Jooste and Van Zyl, 1999). Pro-poor policies that emphasise income generation perform best and have high potential for 'sustained and broad-based poverty reduction' when they are anchored on strategies that improve market access for the poor (Dorward *et al.*, 2002).

World leaders and the development community have therefore adopted trade as a key strategy for achieving the first Millenium Development Goal of halving extreme poverty by 2015 (UNDP, 2003; World Bank, 2003). The emphasis is now more on openness to trade, rather than trade *per se*, as an 'important element of sound economic policy towards alleviation of poverty' (Van Schalkwyk and Jooste, 2003). It would seem that the reality has finally dawned on the development community that the other alternatives to trade for achieving economic growth in Africa, namely aid and debt relief, are in the long run unsustainable solutions for obvious reasons (UNDP, 2003). The Southern African Development Community (SADC), which emerged in its present form following the Windhoek Treaty of 1992, has similarly adopted the enhancement of intra-SADC trade as an article of faith (Chauvin and Gaulier, 2002; Hess, 2002; Mafusire, 2002, among others). This regional body recognises the 'central role for trade within southern Africa, as well as trade between the region and the rest of the world...' (Jooste and Van Zyl, 1999).

1.2.2 Role of agriculture in African economic development and recovery

There is a strong belief that agriculture will drive much of the economic growth on the African continent over the medium to long-term (Eicher, 1999; NEPAD, 2003). Having now recovered from what Eicher (1999) described as 'industrial fundamentalism' of the immediate post-Independence era, African leaders increasingly see that '...Africa will be for many generations, primarily a producer of agricultural and other primary products' (Eicher, 1999, quoting a prominent African leader). The past food and humanitarian crises in the Southern African region serve as a powerful reminder that ignoring agricultural development only accelerates the pauperisation of a region with an unemployment rate of 78% (Van Schalkwyk, 2003). With about 40% unemployment rate and the poor conditions in many 'townships' and 'locations', South Africa has many of the features of a typical dualistic economy. The Cotonou Agreement that provides for non-reciprocal and duty-free access to the European Union markets for poor countries of Africa, Caribbean and the Pacific (ACP), does not classify South Africa as a poor country (European Union, 2000; Makhura and Mokoena, 2003). The continent-wide picture of the incidence of chronic hunger, estimated at some 200 million during 1997-1999 (NEPAD, 2003), is evidently worse for southern Africa where, according to UNDP (2003), '...a set of interlocking problems of unprecedented proportions and complexity, namely, the world's highest levels of HIV/AIDS, food insecurity, and weakened capacities...' are at play.

1.2.3 Impact of market and agrarian reforms

In the words of the Executive Secretary of the SADC (2003), '...it is a question of management'. But he recognises that the question goes deeper than that and adds: 'we are looking at irrigation...at other issues, such as agrarian reform, how we can bring extension services to the rural areas, how we can use information technology'. Many countries in Africa have since the 1990s been dismantling government controls and converting to market-based food systems, believing that market reforms would enhance farm profitability through their positive effects on prices, investment levels, and commercialisation (Jayne *et al.*, 1997; Thorbecke, 2000). In fact, the need for such agrarian reforms, including commercialisation of the smallholder production systems, has received considerable attention from governments and development organisations, including the SADC (FANRPAN, 2003). But the results of the reform programmes have been mixed and frequently inconsistent with the expected increases in productivity. It is now being realised that the sectoral reform prescriptions of the past have, in many cases, been based upon superficial knowledge of the prevailing economic institutions and how they affect economic outcome in particular economies. There is also an emerging general consensus that future productivity growth within the evolving market economies in Africa will require closer attention to the institutional details of the system – i.e. going beyond generalisations that property rights, market rules, and exchange mechanisms need to be identified and worked out, to actually conducting pragmatic applied research on the specific kinds of property rights, rules, and exchange arrangements that would most contribute to economic development under particular circumstances. The process of globalisation has a major impact on the development of institutions and should be taken into account in future research.

1.2.4 Role of institutions

It has been argued that trade reform induces institutional change which in turn facilitates economic growth (Roe, 2003). From numerous studies on the subject of institutional economics (e.g. Dorward, 2001; Dorward *et al.*, 2002; Hall and Soskice, 2001; North, 1990) it is clear that institutions, defined broadly as 'rules of the game', encompassing such elements as transactions costs and risks, information flows, property rights, etc., can enhance or inhibit market access and market development. According to MacFarlan *et al.* (2003), Rodrik (1998) and Rodrik *et al.* (2002), institutions can be viewed in terms of their 'market-creating' or 'market-inhibiting' attributes, to the extent that such issues as property rights and rule of law influence the participation of persons and groups in the economic lives of their societies. Van Schalkwyk and Jooste (2003) have drawn from studies in South Africa and the region to suggest that institutions play an important role in promoting or hindering economic performance in general, and market access in particular.

1.3 Relevance of study to development in South Africa

The dualism that characterised the apartheid-era in South African agriculture remains till today despite more than a decade of reforms. The rural smallholders' impoverishment has not abated probably as a result of inadequate economic infrastructures and lack of access to markets where surplus output can translate into enhanced profitability for the poor rural farmer. The Government has declared that the second post-apartheid decade is devoted to tackling this problem. This vision is consistent with the goals defined under various government programmes, including the Reconstruction and Development Programme (RDP), the Integrated and Sustainable Rural Development Strategy (ISRDS), Black Economic Empowerment (BEE), the Broadening Access to Agriculture Thrust (BATAT) initiative, and others. A sound analysis of these initiatives is needed to obtain empirical data and comprehensive information on the extent and nature of the constraints and opportunities as a guide to more effective programming of development strategies to attain the goals of efficiency, equity and sustainability defined in the Strategic Plan for Agriculture (SPA).

Government and donor policies converge on the need for more effective local development efforts to integrate fully South Africa's smallholders into the mainstream development process. This study coincides with renewed concern at the policy level in the country to address the problems arising from market liberalisation which may have had considerable impact on information flows and a range of ancillary marketing functions including storage, transportation, financial aspects, food processing, etc. The results of this study will make an important contribution towards addressing the fallouts of the marketing reforms on a sustained basis, especially as they relate to the special circumstances of the emerging farmers and the smallholders in the former homelands of South Africa.

1.4 Objectives, research questions and research methodology

This book is based on a systematic information gathering process and analytical framework that were aligned to the thematic interests of the South Africa-Netherlands Research Programme on Alternatives in Development (SANPAD) during the funding cycle of 2004. Nonetheless, the issues remain valid today. The objectives and research questions, as well as the study methodology followed by the various components are elaborated in the following sections.

1.4.1 Objectives and research questions

The smallholder agricultural sector is constrained both by its history of past deprivation and focus on more macro-level reform processes which largely by-pass them. Although the existence of smallholders has been recognised, studies and policy actions generally excluded them, but nowadays more emphasis is being given to access of the emerging and small-scale farmers to institutions, which can improve their livelihoods. In this study, we wanted to ascertain the specific constraints to the development of the smallholder.

The study aims to investigate the institutional and technical constraints to smallholder agriculture in selected areas of South Africa. To what extent can the small-scale farmers share in the expected gains of integration into the national or international markets and what institutional and other reforms are necessary to enhance their effective and profitable participation in the regional economy? This is the central research question that informed the entire research design and the conduct of the research process. The specific research objectives, narrowly linked to the research questions, are partitioned into three levels, namely the micro, meso, and macro levels as follows:

Micro-level research objectives:
- Identify key production and marketing constraints faced by these smallholders.
- Investigate the degree of participation of these smallholders in both input and output markets.

Meso-level research objectives:
- Determine the kind of farmer-based structures and institutions that are needed to empower smallholder farmers to address their constraints.
- Investigate the feasibility of governance systems that can be used in the supply chains of farm commodities produced by smallholders.

Macro-level research objectives:
- Recommend to stakeholders and policy makers how to improve the institutional and policy environment of smallholders.

1.4.2 Research methodology

In general, there are three main research designs in small farmer studies, namely welfare optimisation of producers, analysis of market access of farmers, which involves investigation of market channels and supply chains, and institutional analysis aimed at improving the institutional environment for smallholder agriculture focusing on infrastructure, markets, and human capital aspects. The present study embraced the second and the third strands and therefore employed a variety of methodologies to answer the central research question. A summary of the methodologies used, in relation to the research objectives, is provided below and subsequently elaborated in a separate section at the end of this introductory chapter.

Methodologies related to the micro-level objectives

For this set of objectives, the characteristics of the smallholder farm sector, and present information about their dominant production practices, constraints and opportunities were obtained. Specifically, the methodologies were selected to:
- analyse results of previously conducted studies;
- conduct discussions with groups of key informants and/or a representative set of smallholders;
- conduct surveys to clarify issues that cannot be obtained or solved through analysis of secondary data or group discussions.

Methodologies related to the meso-level objectives

A range of methodologies were employed to achieve this set of objectives, which are described briefly:
- determine the kind of farmer-based structures and institutions that are needed to empower smallholder farmers to address their constraints by reviewing the literature in this respect;
- investigate the feasibility of governance systems that can be used in the supply chains of commodities produced by smallholders.

Methodologies related to the macro-level objectives

Under this category, descriptive statistics were employed to specifically:
- analyse smallholders' access to supply chains of commodities which enter national or international trade;

- recommend to policy makers on how to improve the institutional policy environment of smallholders.

1.5 Description of selected study areas

In the original design, three provinces were identified for the study: Eastern Cape, Free State, and North West Provinces. Within the Eastern Cape, the focus was the former Ciskei homeland where the Nkonkobe Local Municipality was selected. Within the Free State, the study focused on Qwaqwa area, another former homeland/self-governing territory during the Apartheid era. In the North West Province, the study area was the Taung area. Although the book embraces a much wider context than the more restricted geographical setting identified in the original design, this description will focus on the original three provinces since they are representative of the prevailing conditions.

1.5.1 Eastern Cape Province – Ciskei and Nkonkobe

Introduction

The Nkonkobe Municipality of the Amathole District covers a relatively large area, which is predominantly in the former Ciskei Homeland but also includes a portion of the former Cape Province. The towns of Alice, Seymour, and Middledrift, in the former Ciskei, and Fort Beaufort and the hamlet of Hogsback, in the former Cape Province, make up the municipality. The agricultural activities as well as the intensity of agricultural practices vary quite noticeably from one part of the municipality to the other. It also includes the white commercial sector, which concentrates on citrus and beef production. Agriculture in the former Ciskei portion consists of the traditional agricultural sector farming on communal land, emerging commercial citrus farmers and irrigation schemes. This makes it an interesting area to study from the point of view of the institutional and technical constraints to the various forms of agriculture. Another reason for the choice of the Nkonkobe Municipality is the fact that a large body of research that has been conducted in the area. The Faculty of Science and Agriculture and the Agricultural and Rural Development Research Institute (ARDRI) of the University of Fort Hare have been active in the local municipality through Masters and Doctoral theses and research funded by local and international organisations.

Profile of farmers

Studies carried out at the University of Fort Hare have provided characterisations of the smallholder farmer in the Nkonkobe Local Municipality and the former Ciskei homeland. The most recent studies include those by Brown (2000), Dlova (2001), Eiseb (2000), and Monde (2003) which characterised small-scale farmers of the Municipality over a broad range of settings. The study by Eiseb (2000) specifically examined small-scale farmers in terms of their physical and human resource capabilities and constraints, while Brown (2000) focused on the reconciliation of the dominant stereotypes of black farmers against the realities on the ground. Monde (2003) reported on the food security status of households in two villages of the former Ciskei homeland. Overall, the studies confirmed that the majority of small-scale farmers in the municipality generally had limited education, were poorly resourced in terms of capital endowment and land, and lacked sufficient access to institutional credit or technical support that would enable them expand production and improve their livelihoods. It was clear that the effects of past discriminatory policies were still in evidence and that recent policies to achieve greater inclusivity had had limited success.

Marketing of agricultural products

The marketing of agricultural products in the Nkonkobe Municipality follows the same pattern that has been observed in many other smallholder environments where small-scale producers deliver a meager output to diverse formal and informal outlets, often in the absence of a systematic coordinating mechanism. The product range can be extensive even if limited to the dominant vegetable and food grain staples of cabbages, tomatoes, onions, butternut, peas, beetroot, potatoes, watermelon, pumpkins, maize, sugar beans, peas, spinach and other leafy vegetables. Direct sales to consumers at the farm gate seem quite frequent although a few producers truck produce to the towns of Alice and Fort Beaufort. It is also common to see the retailers going into the villages to purchase the produce from producers for sale in the towns. Undoubtedly, small farmers in the municipality face the serious challenges of poor infrastructure in delivering produce to markets where profitable sales can be undertaken. It was clear from preliminary investigations that this would be a major focus of the study to identify such constraints that the smallholders face in the marketing of produce that could be factored into recommendations about unlocking markets to smallholders.

1.5.2 Free State Province – Qwaqwa area

Introduction

Qwaqwa is a former homeland area, which was granted self governance on 1 November, 1974. On 27 April, 1994 it was reunited with South Africa, together with the nine other former homelands. The area is situated in the eastern part of the Free State, bordering KwaZulu-Natal and Lesotho. It is home to more than 180,000 Sotho-speaking South Africans. It covers an area of about 655 km^2. During the period 1986 to 1994, about 70,000 ha were identified and developed for resettlement of black farmers.

Profile of Qwaqwa farmers

A study conducted by Jordaan and Jooste (2003) provided a thorough descriptive profile of the emerging farmers of Qwaqwa. The farmers generally have had little formal schooling with, on average, 15% of the farmers without any school qualification, whilst 65% were qualified at primary school level. Only 6% have post matric qualifications. A large majority of the farmers are above 50 years old. Approximately 20% are between 35 and 50, and only 2% are younger than 35. Jordaan and Jooste (2003) found that only 12% of the farmers do not have their own title deeds. They lease their land from the Department of Land Affairs, but expressed their desire to purchase the land they lease. The fact that 88% of the farmers actually own the land they farm could increase the commitment to find solutions to their problems to secure their ownership rights. In general, most of the farmers in Qwaqwa have also non-farm income, which is used for agricultural enterprises.

Marketing of agricultural produce

In Qwaqwa the main crops produced are maize, wheat, dry beans, soybeans and sunflower seed. Beef and milk production are the most important livestock enterprises. The major challenges farmers face are successful marketing of their products.

In terms of maize, the nearest marketing point for Qwaqwa farmers is Genfood, with their Maluti maize mills, situated in Phuthadithjaba. The maximum distance from farms to the maize mills is 40 km. The Maluti maize mill's demand for maize is very high and it has under-utilised capacity. This provides great potential for expanding the maize production in the area. Three silos in the area are owned by OTK and are situated in Harrismith and Afrikaskop. None of these silos is further than 40 km from the Qwaqwa farms. OTK and VKB are the main buyers of wheat in the area. Silos equipped with drying capacity are also located in Harrismith and Afrikaskop. Dry beans are a risky enterprise, but the high income per unit is high enough to create an incentive to produce it. OTK, VKB and many cash buyers, especially from KwaZulu-Natal, are the main buyers of dry beans. Soy beans and sunflower seeds are enterprises with much room for improvement. There are market possibilities in the whole of the Eastern Free State through OTK and VKB.

Marketing of milk seemed to face more problems than realised. Seasonal production is a serious concern, for milk production is very low in the cold and dry winter months. There is also a lack of cooling facilities for the farmers, and traveling distance is very far to the delivering points. Three cooling and collection points have been established at strategic points. Farmers deliver their daily production at the delivery point nearest to their farms, and tankers from Thaba Dairies collect the milk. Farmers get paid according to the quality of their milk. In the past, Thaba dairies used to buy milk from AgriEco, of which 50% of the shares were kept by the dairy farmers in Qwaqwa. The revitalising of this cooperative can open great potential for milk production in Qwaqwa (Free State Government, 2004).

The food retail environment is rather well developed, with two cash and carry wholesalers, three large chain supermarkets, two non-chain supermarkets, multiple small supermarkets and numerous fruit and vegetable stalls, tuck shops and spaza shops. There are also a few small butcheries, bakeries, catering services, shebeens, taverns and restaurants. A large fruit and vegetable market also exists and provides an avenue for small producers to sell small volumes and for consumers to buy at lower prices (Botha, 2005).

1.5.3 North West Province – Taung area

Background

The North West Province is 116,320 square kilometers in size, consisting of four core district municipalities of Bojanala, Central, Southern and Bophirima. The fifth district is designated a Presidential Node that embraces parts of the Northern Cape Province. The province boasts of agricultural, mining and tourism opportunities. Crops like maize, tobacco, paprika, etc. are grown as cash crops; mining of platinum, diamond and gold bring in substantial wealth, and tourist attractions like the Sun City and Pillanesburg Nature Reserve have become synonymous with the provincial success story. However, the predominantly rural black population, more than 60% of the total population of 3.6 million, are still living below the poverty line.

Poverty pockets

A recent poverty assessment by the National Development Agency (NDA) identified at least two poverty pockets from each district municipality which include the following: Molopo, Naledi, Greater Taung, and Lekwa – Teemane in Bophirima; Setla Kgobi, Tswaing and Ditsobola in Central; Madibeng, Moretele, and Mankwe – Madikwe in Bojanala; with Maquassi Hills and Ventersdorp featuring highly in the Southern District.

Elements for development of agricultural production and marketing

North West Province is an important livestock producer, emphasis being on cattle. According to the Provincial Government, agricultural production and marketing face serious difficulties as a result of several factors. One key problem relates to the poor state of the province's physical infrastructure and the fact that the agricultural districts are remote and often inaccessible. An additional complication which has more serious implications for marketing of agricultural products is the question of subsidised imports (Khasu, 2000). In relation to this, the government pins a lot of hope on the outcome of trade negotiations within the frameworks of the World Trade Organization (WTO) and the European Union (Khasu, 2000). To address provincial developmental challenges, three main economic drivers have been identified as agriculture, mining and tourism, with agriculture obviously seen as a prime mover. From these drivers, several potential projects have been highlighted. The provincial Departments of Agriculture, Conservation and Environment, Economic Development and Tourism and Social Services, are already involved in a variety of projects that have been identified by both municipalities and stakeholders. In 2007, the department further strengthened its Economics Unit to focus on Marketing by establishing an Agricultural Marketing Services sub-directorate which was charged to: conduct agricultural market infrastructure needs assessment in the districts, explore opportunities for linking farmers to markets, facilitate the flow of agricultural market information to farmers, including local and international trade updates and futures through forging links with the SAFEX.

Unemployment is very high in the North West with rates being put at as high as 48.6% in 2005 which makes the province better than only two provinces in the country, namely Limpopo and the Eastern Cape. Again, at 48.6%, the rate of unemployment in North West province is higher than the national average rate of 41.5%. It is estimated that the number of households in the lowest income groups has increased from 13% to 22% between 1996 and 2001[1]. Indications are that this trend is continuing. A report released in 2005 by the South African Brewery in relation to its development activities in the province, estimated that unemployment rates in Taung stood at about 80%, with monthly average incomes at about R200 per capita. On a broad provincial level, it is estimated that more than 60% of households earn less than R1500 per month, and automatically qualify for government support programmes. Examples of the most-affected areas include Phokwani, Huhudi, Schwezer – Renecke and Delareyville (predominantly in Bophirima and Kgalagadi District Councils).

1.6 Data collection and data analysis

In recent years, our general knowledge about supply chains and marketing channels (e.g. Ruben *et al.*, 2007; Van Tilburg *et al.*, 2007), has increased, but reality varies with product and location. For instance, knowledge about the size and nature of existing market types (i.e. spot markets such as assembly markets, fresh product markets, wholesale markets, auctions, etc.) serving the smallholder sector remains incomplete. It is important to know how prices are determined in these locations and what the implications are for transaction costs and invariably the feasibility of trade. It is also desirable within this context to determine the extent of coordination among markets, where they already exist, and what mechanisms bring about such coordination. To resolve the informational gap, a survey within the selected provinces and study municipalities/areas took place, to obtain insight into the existing situation on the ground. Prior to the surveys, a series of case studies had been conducted to gain deeper insights into the operations of small-scale or emerging farms and the constraints they face in production and marketing. These case studies provided one basis for sampling for the field survey phase of this research. This knowledge has also been useful in affording fuller understanding of the nature of the problem to be addressed and how the population is distributed around such issues. As might be expected, this was handy in defining the recommendation domain and also articulating the broader context essential for setting the

[1] Background of North West Province: http://www.discoverytoursafrica.com/north-west.htm p. 4, 5/4/2004.

policy agenda for unlocking markets for smallholders which is discussed in the concluding chapter of this book.

1.7 Organisation of the book

The book is organised in 11 chapters. The first part (Chapter 1) sets the scene by reiterating the theme and the theoretical case for the study. The single chapter making up this section provides an overview of the history and evolutionary context of market policies in the country and traces the changes in the institutional environment for smallholder development in the country. By so doing, the chapter revisits and restates the debate and the initial assumptions on which the strong case was made for the study. The second part (Chapters 2-6) turns attention to the specific issue of constraints analysis with particular reference to the depressed areas, or former homelands. The third part (Chapters 7-10) of the book focuses on the meso-level to see how constraints operating at the micro-level influence development-relevant coordination of the national and international food systems. In this regard, the issues of supply chain governance, food retailing and credit accessibility, are examined in relation to their impact on smallholder and agricultural development. Chapter 11 draws on the lessons learnt through the various chapters of the book, bringing the theory and learning together resulting in recommendations in relation to the stakeholder groups addressed.

References

Aerni, P., 2008. A new approach to deal with the global food crisis. African Technology Development Forum 5: 16-31.

Bienabe, E., C. Coronel, J-F. Le Coq and L. Liagre, 2004. Linking smallholders to markets: lessons learned from literature review and analytical review of selected projects. World Bank/CIRAD/IRAM, Washington, DC, USA.

Botha, L., 2005. Can we get them there? A case of commercializing arable farming at Rust de Winter farms of Limpopo province, South Africa. Available at: http://www.icra-edu.org/objects/public_eng/SA2005.pdf.

Brown, L.R., 2000. In search of a systems model for the decision making behaviour of first-generation black commercial farmers in the Border-Kei region of the Eastern Cape province. PhD thesis, University of Fort Hare, Alice, South Africa.

Chauvin, S. and G. Gaulier, 2002. Prospects for increasing trade among SADC countries. Paper presented at the 2002 Annual Forum on Trade and Industrial Policy Strategies, Glenburn Lodge, Muldersdrift, South Africa.

Church, N., 2005. Why our food is so dependent on oil. Available at: http://Countercurrents.org.

Department of Provincial & Local Government, 2003. Budget Review Speech by N.G.W. Botha, M.P., Deputy Minister. Provincial and Local Government at National Assembly on 12 June 2003.

Department of Social Development, 2010. Budget speeches and analysis of budget debates. Department of social development, Pretoria, South Africa.

Dlova, M.R., 2001. Agricultural production in the Hertzog Agricultural Cooperative of the Seymour district in the Eastern Cape, South Africa. M Agric dissertation, University of Fort Hare, Alice, South Africa.

Dorward, A., 2001. The effects of transaction costs, power and risk on contractual arrangements: a conceptual framework for quantitative analysis. Journal of Agricultural Economics 52: 59-74.

Dorward, A., N.D. Poole, J. Morrison, J. Kydd and I. Urey, 2002. Critical Linkages: livelihoods, markets and institutions. ADU Working Paper 02/03. Available at: http://ageconsearch.umn.edu/handle/10919.

Eicher, C.K. and M. Rukuni, 1996. Reflections on agrarian reform and capacity building in South Africa. Staff Paper No. 96-3, Department of Agricultural Economics, Michigan State University, MI, USA.

Eicher, C.K., 1999. Institutions and the African farmer. Issues in Agriculture 14, Consultative Group on International Agricultural Research (CGIAR), Washington, DC, USA.

Eiseb, M.M., 2000. Characterisation of small-scale farmers in the Keiskammahoek magisterial district of Eastern Cape. MSc Agric dissertation, University of Fort Hare, Alice, South Africa.

European Union, 2000. ACP-EU partnership agreement signed in Cotonou on 23 June 2000, Directorate General for Development of the European Commission, Brussels, Belgium.

Food Agriculture and Natural Resources Policy Analysis Network (FANRPAN), 2003. Equipping SADC member states for effective participation in multi-lateral agricultural trade negotiations. Policy Discussion Paper No. Pan 2/03, FANRPAN, Harare, Kenya.

Food and Agriculture Organization of the United Nations (FAO), 2008. Soaring food prices: facts, perspectives, impacts and actions required. Paper prepared for the High-Level Conference on World Food Security, 3-5 June 2008. Food and Agriculture Organization, Rome, Italy. Available at: ftp://ftp.fao.org/docrep/fao/.

Free State Government, 2004. Accelerating economic growth and development in the Free State: framework for a responsive partnership approach towards sustainable economic development in the Free State. Mimeo. Available at http://www.fs.gov.za/INFORMATION.

Gabre-Madhin, E., 2006. Making markets work and work for the poor. International Food Policy Research Institute (IFPRI), Washington, DC, USA.

Government of South Africa, 2004. Budget Speech by Minister of Finance Mr. Trevor Manuel. Department of Finance, Pretoria, South Africa.

Green, B.M., 1978. Eating oil – energy use in food production. Westview Press, Boulder, CO, USA.

Hall, P.A. and Soskice, D., 2001. An introduction to varieties of capitalism. Available at: http://kisi.deu.edu.tr/muge.tunaer/VoC.pdf.

Hess, S., 2002. Economic geography and the implications of a free trade area within SADC. Paper presented at the 2002 Annual Forum on Trade and Industrial Policy Strategies, Glenburn Lodge, Muldersdrift, South Africa.

International Fund for Agricultural Development (IFAD), 2003. Promoting market access for the rural poor in order to achieve the Millenium Development Goals. Roundtable discussion paper for the twenty-fifth anniversary session of IFAD's governing council, IFAD, Rome, Italy.

Jayne, T.S., J.D. Shaffer, J.M. Staatz and T. Reardon, 1997. Improving the impact of market reform on agricultural productivity in Africa: how institutional design makes a difference. Michigan State University International Development Working Paper No. 66, Michigan State University, East Lansing, MI, USA. Available at: http://ideas.repec.org/p/ags/midiwp/54684.html.

Jooste, A. and J. Van Zyl, 1999. Regional agricultural trade and changing comparative advantage in South Africa. Technical paper No. 94, SD Publication Series, United States Agency for International Development (USAID), Washington, DC, USA.

Jordaan, A.J. and A. Jooste, 2003. Strategies for the support of successful land reform: a case study of Qwaqwa emerging commercial farmers. South African Journal of Agricultural Extension 32: 1-14.

Khasu, M.J., 2000. Budget speech 2000/2001. Member of Parliament and Member of Executive Council for Agriculture, Conservation and Environment, North West Province, Soutn Africa, 11 April 2000.

Krugman, P., 2008. The increasing returns revolution in trade and geography. *Prize Lecture*, Nobel prize in economics, Swedish Academy, Stockholm, Sweden. Available at: http://nobelprize.org/nobel_prizes/economics/laureates/2008.

Leamer, E.E., 1995. The Heckscher-Ohlin Model in theory and practice. Princeton Studies in International Finance No 77, Princeton University, Princeton, NJ, USA.

Lyne, M.C., 1996. Transforming developing agriculture: establishing a basis for growth. Agrekon 35: 188-192.

MacFarlan, M., H. Edison and N. Spatafora, 2003. World economic outlook. The International Monetary Fund (IMF), Washington, DC, USA.

Mafusire, A., 2002. SADC trade: challenges and opportunities to the regional countries. Paper Presented at the 2002 Annual Forum on Trade and Industrial Policy Strategies, Glenburn Lodge, Muldersdrift, South Africa.

Makhura, M. and M. Mokoena, 2003. Market access for small-scale farmers in South Africa. In: L. Nieuwoudt and J. Groenewald (eds.) The challenges of change: agriculture, land and the South African economy. University of Natal Press, Natal, South Africa, pp. 137-148.

Masters, W.A., 2008. Beyond the food crisis: trade, aid and innovation in African agriculture. African Technology Development Forum 5: 3-13.

Mbongwa, M., R. Van den Brink and J. Van Zyl, 1996. Evolution of the agrarian structure in South Africa. In: J. Van Zyl, J. Kirsten and H.P. Binswanger (eds.) Agricultural land reform in South Africa. Oxford University Press, Cape Town, South Africa, pp. 36-63

Minot, N. and R.V. Hill, 2007. Developing and connecting markets for poor farmers. 2020 focus brief on the world's poor and hungry people, International Food Policy Research Institute (IFPRI), Washington, DC, USA.

Mokoena, M.R., 2002. An overview of South Africa's international and regional trade relations: implications on the agrarian reform. Working document, The Human Science Research Council (HSRC), Pretoria, South Africa.

Monde, N., 2003. Household food security in rural areas of central Eastern Cape: the case of Guquka in Victoria East and Koloni in Middledrift districts. PhD thesis, University of Fort Hare, Alice, South Africa.

National Department of Agriculture, 2001. The strategic plan for South African agriculture, Directorate Agricultural Information Services, Pretoria, South Africa.

Ndirangu, N., 2002. Africa worse-off after agreement on agriculture. Paper presented at the Roundtable on Food and Trade: the WTO Development Change, November 4-5, 2002, Ottawa-Ontario, Canada.

New Partnership for Africa's Development (NEPAD), 2003. Comprehensive Africa agriculture development programme. New Partnership for Africa's Development, Midrand, South Africa.

North, C., 1990. Institutions, institutional change and economic performance. Cambridge University Press, Cambridge, UK.

Nyamande-Pitso, A., 2001. The experience of black business in the import/export market. Paper presented at the 8th Annual Agriculture Management Conference, October 30, 2001, Midrand-Johannesburg, South Africa.

Obi, A., 2006. Trends in agricultural land prices. PhD Thesis, University of the Free State, Bloemfontein, South Africa.

Ortmann, G. and C. Machethe, 2003. Problems and opportunities in South African Agriculture. In: L. Nieuwoudt and J. Groenewald (eds.) The challenges of change: agriculture, land and the South African economy, University of Natal Press, Natal, South Africa, pp. 47-62.

Ricardo, D., 1817. The principles of political economy & taxation. John Murray, London, UK.

Rodrik, D., 1998. Trade policy and economic performance in Sub-Saharan Africa. NBER Working Paper Series No. 6562, National Bureau of Economic Research, MA, USA.

Rodrik, D., 2002. Trade policy reform as institutional reform. In: B. Hoekman, A. Mattoo, and P. English (eds.) Development, trade and the WTO: a handbook, The World Bank, Washington, DC, USA, pp. 3-10.

Rodrik, D., A. Subramanian and F. Trebbi, 2002. Institutions rule: the primacy of institutions over geography and integration in economic development. NBER Working Paper, No. w9305, National Bureau of Economic Research, MA, USA.

Roe, T, 2003. Markets, trade and the role of institutions in African development. Paper presented at the pre-IAAE Conference, 13-14 August, 2003, at the President Hotel, Bloemfontein, South Africa.

Ruben, R., M. Van boekel, A. Van Tilburg and J. Trienekens, 2007. Tropical food chains: governance regimes for quality management. Wageningen Academic Publishers, Wageningen, the Netherlands.

South African Development Community (SADC), 2003. Press briefing: end-of-year briefing on regional developments by SADC executive secretary Dr. Prega Ramasamy. 9 December, 2003.

Southgate, D., D. Graham and L. Tweeten, 2007. The world food economy. Wiley-Blackwell, Malden, MA, USA.

Southgate, D., 2010. Long-term trends in population, food demand and commodity prices. Ohio State University, OH, USA.

Summers, L.H., 1992. Keynote address: knowledge for effective action. Proceedings of the World Bank Annual Conference on Development Economics 1991, International Bank for Reconstruction and Development/The World Bank, Washington, DC, USA.

Summers, L.H. and V. Thomas, 1993. Recent lessons of development. The World Bank Research Observer 8: 241-254.

Thorbecke, E., 2000. Agricultural markets beyond liberalization: the role of the state. In: A. Van Tilburg, H.A.J. Moll and A. Kuyvenhoven (eds.) Agricultural markets beyond liberalization, Kluwer Academic Press, Dordrecht, the Netherlands, pp.19-53.

United Nations Development Programme (UNDP), 2003a. The capacity initiative for Southern Africa – meeting the challenge of collapsing capacity in Southern Africa in times of HIV/AIDS. Draft Internal UNDP Discussion Document 10 July, 2003. UNDP, New York, NY, USA.

United Nations Development Programme (UNDP), 2003b. Human development report 2003 – Millenium Development Goals: a compact among nations to end human poverty. UNDP, New York, NY, USA.

Van Schalkwyk, H.D., 1995. Modeling South African agricultural land prices. PhD thesis, University of Pretoria, Pretoria, South Africa.

Van Schalkwyk, H.D., 2003. Intra-regional trade and economic development in Southern Africa. Paper presented at the pre-IAAE Conference, 13-14 August, 2003, at the President Hotel, Bloemfontein, South Africa.

Van Schalkwyk, H.D. and A. Jooste, 2003. The role of trade reform in growth and poverty reduction. Paper presented at the pre-IAAE Conference, President Hotel, Bloemfontein, South Africa.

Van Schalkwyk, H.D., J. Groenewald and A. Jooste, 2003. Agricultural marketing in South Africa. In: L. Nieuwoudt and J. Groenewald (eds.) The challenges of change: agriculture, land and the South African economy. University of Natal Press, Natal, South Africa, pp. 119-136.

Van Tilburg, A., 2003. Framework to assess 'worldwide' the performance of food marketing systems. Paper presented at the Brown Bag Seminar of the Department of Agricultural Economics, November 11, 2003, Michigan State University, East Lansing, MI, USA.

Van Tilburg, A., J. Trienekens, R. Ruben and M. Van Boekel, 2007. Governance for quality management in tropical food chains. Journal on Chain and Network Science 7: 1-9.

Van Zyl, J. and H.P. Binswanger, 1996. Market-assisted rural land reform: how will it work? In: J. Van Zyl, J. Kirsten and H.P. Binswanger (eds.) Agricultural land reform in South Africa, Oxford University Press, Cape Town, South Africa, pp. 3-17.

Wakeford, J.J., 2006. The impact of oil price shocks on the South African macroeconomy: history and prospects. South African Reserve Bank Conference, Johannesburg, South Africa.

World Bank, 2002. Empowerment and poverty reduction – a sourcebook. International Bank for Reconstruction and Development/The World Bank, Washington, DC, USA.

World Bank, 2003. World bank development report 2003 – Sustainable development in a dynamic: transforming institutions, growth, and quality of life. International Bank for Reconstruction and Development/The World Bank, Washington, DC, USA.

World Bank, 2007. World development report 2008: agriculture for development. International Bank for Reconstruction and Development/The World Bank, Washington, DC, USA.

2. Strategies to improve smallholders' market access

Aad van Tilburg and Herman D. van Schalkwyk

2.1 Introduction

2.1.1 Market access

Smallholders, especially in less developed countries, have encountered several challenges in gaining access to markets. Market access includes the ability to obtain necessary farm inputs and farm services, and the ability to deliver farm products to buyers. Market access was less of a problem in the era of the marketing boards, roughly from 1940 to 1990, when a parastatal organisation – the marketing board – tended to provide essential farm inputs such as seed, fertilisers and ploughing services, farm services such as extension and credit, and output market services such as collection of the harvest, quality assessment and buying. Marketing boards tended to issue pan-seasonal and/or pan-territorial product prices and purchased from farmers and traders at several central locations. The consequence of this approach was that the decisions made by producers, processors, transporters, traders and consumers were not fully guided by free market principles and prices as indirect subsidies were involved. Marketing boards were dissolved in the 'eighties' and 'nineties' in the majority of developing countries because their activities, as a rule, appeared not to be economically sustainable and consequently a heavy burden on the national government's budget. Consequently, smallholders were suddenly deprived of a supportive institutional marketing structure. This was also the case in South Africa with the repeal of the Marketing Act of 1968 and the implementation of the Marketing of Agricultural Products Act of 1996 (e.g. Van Schalkwyk *et al.*, 2003).

The abolishment of parastatals in South Africa in the 'nineties' was a serious drawback notably for smallholders – and among them – emerging farmers seeking access to markets while they still did not have sufficient experience to operate in a competitive free market environment. This meant that emerging farmers operating in the former homelands of South Africa were deprived of substantial support. Although quite a number of emerging farmers were given access to land, they did not usually receive a title deed, thus preventing them from using their land as collateral for both investment and working capital. Farmers, especially those operating in the former homelands, generally encountered high transaction and transport costs to access markets. Exceptions were those cases where agribusiness companies concluded contracts with smallholders in order to procure essential commodities such as milk, sugarcane or barley (e.g. Tregurtha and Vink, 1999).

2.2.1 The societal context

Problems and opportunities of smallholders were summarised in the Strategic Plan for South African Agriculture (Department of Agriculture, 2001). Aspects of this plan, relevant to this chapter, concern institutional reform affecting the governance of agriculture, the skewed participation of smallholders in society, the challenge of unlocking the untapped creative energy of people, improvements in the support and delivery systems for smallholders and sustainable management of essential natural resources. Several strategies to deal with these challenges were presented in this strategic plan: an equitable access and participation strategy; a land reform programme; a programme of farmer support services; identifying emerging farmers from historically disadvantaged groups; the initiation of innovative development programmes for farmers operating on communal land; strategies for improved supply-chain performance; generating a possible policy environment; providing risk management services and planning sustainable resource management.

Stakeholders in the agricultural sector have been invited to improve market access by eliminating entry barriers, engaging in collective action, enhancing the transfer of technology, implementing a human resources development plan, improving access to a comprehensive range of rural and financial services including extension and to improve the collaboration and coordination between government institutions, agricultural organisations, non-government organisations (NGO's) and civic associations.

2.1.3 Objective and research queries

The chapter deals with smallholders, notably emerging farmers requiring access to markets of farm inputs and farm services as well as output markets. The aim of the chapter is to suggest strategies for smallholders to obtain or improve market access. To this end the following questions need to be answered:

- What must be done to improve market access for smallholder produce?
- Which stakeholders can contribute to facilitate smallholder access to markets or supply chains and what can be the role of each stakeholder group?
- Which promising alternative strategies can be developed to improve market access?

2.1.4 Illustrative case studies

Several case studies and sources in the literature show how smallholders' market access can be facilitated by strengthening their human, social and economic capital, their countervailing power in the market and by seeking improvements in their physical and institutional environment. The problem areas of smallholders' market access is illustrated with six case studies in South Africa. Major bottlenecks to improve smallholders' market access were the lack of market transparency and bargaining power in the market, lack of group action, lack of necessary farm management skills and insufficient access to resources needed to operate a farm (Table 2.1).

Table 2.1. Main bottlenecks for market access as presented in six case studies.

	Market transparency and bargaining power	Group action	Human and social capital: smallholders' management skills	Access to required resources: land, farm inputs, farm services
1. Market development	✓	✓	✓	✓
2. Sugar cane	✓	✓	✓	✓
3. Agricultural services	✓	✓	✓	✓
4. Wool	✓	✓	✓	✓
5. Rooibos tea	✓	✓	✓	
6. Mentorship		✓	✓	

The main lessons learnt are derived from cases on:

- A marketing design study (Appendix 2.1). Rural development projects that involve beneficiaries in identifying their own choice of alternatives tend to succeed, unlike projects where the project is developed outside the community and the community members are asked to participate. Another lesson learnt from this study was that unlocking Eastern Cape provinces' potential for smallholders or emerging farmers in particular was highly dependent on creating an enabling environment.
- Sugarcane production by smallholders in KwaZulu-Natal (Appendix 2.2). The need to strengthen farm management abilities of SSGs through training and mentorship, the need for investments in machinery by contractors and the need to improve contractor market performance by improved market transparency and more symmetric bargaining power between SSGs and contractors.
- Agricultural Services for smallholders (Appendix 2.3). Potential emerging farmers need to be trained in farm management practices and be embedded in a proper institutional context, in order to qualify for a complete package of supporting services.
- Wool. The Golden Fleece project for smallholders (Appendix 2.4). Improved market access could be obtained through industry initiatives to promote group action and the building of shearing sheds, which also serve as training centres.
- Emerging rooibos tea farmers in the Heiveld community (Appendix 2.5). NGO's played critical roles in facilitating the two projects: social capital has been strengthened; the communities have gained in confidence and enhanced their production, marketing and management skills. NGO's also played a crucial role in linking poor communities to sources of external funding and potential markets.
- The role of mentorship in obtaining market access (Appendix 2.6). The forming of interactive mentorship alliances that are complementary, loosely structured, based on previous experience of smallholders, and not being hindered by complicated rules and

regulations can be successful, provided that they operate in an enabling environment with opportunities for group action by emerging farmers and a level playing field in markets relevant for these farmers

In summary, the lessons learnt from these six case studies are: first, market outlets for smallholder produce need to be developed or improved by stakeholders in the value chain; second, market access for smallholders needs to be improved by increasing market transparency and obtaining more balanced bargaining power in the market through group action; third, to be able to participate in markets, a variety of services (e.g. extension, support to initiate group action, mentorship alliances) are needed to strengthen the human and social capital base of smallholders; and fourth, improving market access for smallholders is very dependent on a suitable enabling institutional environment to obtain the required resources to become a successful farmer.

2.1.5 Organisation of the chapter

Theoretical constructs on the analysis of bottlenecks with respect to market access of smallholders are summarised in Section 2.2. Next, in Section 2.3, experience in how to cope with bottlenecks for smallholder market access reported in the literature, is discussed and combined with insight obtained from the illustrative case studies to generate suggestions to improve market access. Based on the analysis in Section 2.3, a few potential successful strategies to improve market access are developed in Section 2.4. Finally, Section 2.5 summarises the conclusions of the chapter.

2.2 Theoretical constructs on the analysis of market access

Market access of smallholders implies that smallholders can have access to either spot markets or a supply chain that delivers the required market services. The coordination of economic activities between primary producers and consumers in the supply chain can be characterised by its type of governance structure, for example; ownership, contractual or trust-based (e.g. Stern *et al.,* 1996).

2.2.1 Spot market coordination

Spot market coordination of economic activities is governed by markets in which supply and demand is cleared through price discovery. Market prices embody a crucial signalling device directing the decisions of market participants. Theory on the coordination of economic activities through spot markets (e.g. Hill and Ingersent, 1982) has largely been based on the model of perfect competition. Analyses of market performance in reality have been done by comparing actual patterns of competition in a particular spot market with the (theoretical) characteristics of either perfect, workable or contestable competition (e.g. Baumol *et al.,* 1988). Perfect competition – maximising the welfare of buyers and sellers under strict conditions (e.g. Henderson and Quandt, 1980) – is characterised by homogeneous demand, perfect market information, divisible and mobile resources, many buyers and sellers being price takers rather than price setters, and cost-effective transactions. Consequently, conditions of workable competition are quite close to perfect competition, e.g. products are rather homogeneous, there are sufficient buyers and sellers to obtain a level playing field, market transparency is reasonable and barriers to entry or exit are relatively insignificant. The main condition of 'contestable' competition is that market entry and exit are free, resulting in traders ('incumbents') also taking potential competition by new market entrants into account in their strategies (e.g. Van Tilburg, 2010).

2.2.2 Vertical coordination in the supply chain

A hierarchy in guiding economic activities can be obtained through ownership, e.g. in the Chiquita banana supply chain, by means of contracts, or by means of a channel leader's initiative to follow a common marketing plan. A network of economic activities, e.g. amongst relatives or business partners, consists of informal relationships lubricating economic activities between agents. On this basis, common types of coordination in supply chains or marketing channels (Stern *et al.,* 1996, Chapter 6) were labelled as *conventional marketing channels* where spot market competition prevails in each stage of the chain, and *vertical marketing systems* (hierarchies) in which at least two subsequent stages in the chain cooperate through voluntary or contractual coordination, or through networks based on people who trust each other (e.g. Van Tilburg, 2010).

The framework that can be used in this chapter to analyse bottlenecks regarding market access includes (Van Tilburg, 2010) *market structure analysis* in which it is assessed to what extent there is a level playing field in markets, *market integration analysis* to assess the correlations in market price developments in spatially separated markets, *exchange or transaction theory* to assess what affects the outcome of a transaction between trade partners, and *analysis of vertical coordination in the supply chain* to assess opportunities for primary producers to improve their market access and to streamline the flow and quality of products between the stages of primary production and final consumption. This is summarised in Table 2.2.

Based on the preceding discussion on theoretical constructs and the illustrative case studies, the following approach is proposed to analyse bottlenecks for smallholder access to markets or supply chains (Table 2.3). This approach is used in Section 2.3 to discuss each of these bottlenecks, based on both the economic and marketing literature on South Africa.

2.3 Bottlenecks for smallholder market access and the lessons learnt

In this section, insight from the literature on bottlenecks for market access as well as the way bottlenecks for market access have been dealt with in the illustrative case studies are

Table 2.2. Mode of supply chain governance in relation to theoretical constructs (Adapted from Van Tilburg, 2010).

Theoretical construct	Theme	Specification
Spot market coordination		
Organisational economics	industrial organisation: market performance	market structure analysis market integration
Vertical coordination		
Marketing	coordination in the supply chain or distribution channel	ownership contractual network
Organisational economics	coordination through transactions or contracts	contracts in a weak institutional environment

Table 2.3. Bottlenecks to smallholders' market access.

Bottlenecks at what level	Bottlenecks
At farmers level	lack of resources lack of horizontal coordination or group action lack of institutional support
At market level	entry barriers lack of market opportunities
At supply chain level	lack of proper vertical coordination with: • the agribusiness processing sector • the retail sector • the export sector

discussed and used in Section 2.4 to examine what particular stakeholder groups in the supply chain might do to improve market access.

2.3.1 Bottlenecks for market access at smallholders level

Three categories have been serious bottlenecks for smallholders: lack of access to resources, lack of horizontal coordination or group action, and lack of institutional support.

Smallholders' lack of access to resources

Four out of the six illustrative case studies demonstrate the importance of improving smallholders' access to resources. Smallholders in rural areas of South Africa have been subject to high illiteracy rates and reducing these illiteracy levels has been a challenge faced by both public and private stakeholders (Dawson, 2003). School children as well as their parents have been trained in literacy, and there are examples of children assisting their parents in learning to read and write. Training on how product quality concepts and market information has been used to improve the market value of farm produce, has also been made available. In addition, farmers' negotiating skills with respect to the settlement of transactions has been further developed (e.g. Coetzee *et al.*, 2005). Magingxa studied market access in six smallholder irrigation systems in the Eastern Cape, Mpumalanga and Limpopo provinces (Magingxa, 2006; Magingxa *et al.,* 2006). Market access was one of the factors influencing the success potential of smallholder irrigation projects together with farmer skills and membership of a farmer organisation. Other variables affecting the success of smallholder irrigation projects included access to information, training, transport, extension and planning, but nevertheless there was still a need to strengthen smallholders' knowledge, experience and skills to link with new markets. Obtaining sustainable access to markets requires careful planning, coordination and monitoring by all partners in a supply chain (e.g. Bussard and Uhrinova, 2006). Smallholders need to be trained and helped to connect with markets, to learn basic skills and to obtain access to agricultural extension and farm enterprise development services. Nowadays, new ways of communication and distance learning offer new opportunities to improve smallholders' human capital.

The conclusion is that smallholders' access to human and social capital needs further development in order to be able to facilitate access to markets and supply chains. Stakeholders able to play a role in this respect are farmer cooperatives; value chain members such as input suppliers and agribusiness processing companies; sector associations; the trade union NAFU and NGOs. Relevant public stakeholders concern national, provincial or district authorities, e.g. the provincial Departments of Agriculture.

Lack of group action

Each of the six case studies illustrates the importance of group action by smallholders to obtain market access. Smallholders can use either spot markets, or vertical integration by means of contracts, for their market transactions (e.g. Sartorius and Kirsten, 2002). But smallholders require sufficient countervailing power to be able to obtain a fair share of the value added in the supply chain. A member-dominated cooperative can enable smallholders to reap both economies of scale while increasing their countervailing power. Lack of human, social and economic capital among potential members could be restrictive in starting a successful cooperative (e.g. Christie, 2001). The main reasons for forming smallholder cooperatives in two communal areas in KwaZulu-Natal, were market failure, missing services, assurance of input supply, income-generating opportunities, increase in countervailing power and reduction of transaction costs. Both strength of community leadership and motivation of members appeared to be weak points (Ortmann and King, 2007). Holloway *et al.* (2000) reported, in a study on milk marketing by small-scale farmers in the east-African highlands, that cooperative selling institutions are potential catalysts for reducing transaction costs, and can stimulate market entry and promote growth in rural communities. Several producer cooperatives or outgrowers' schemes were useful instruments in overcoming barriers to market access. But, in several studies, considerable ignorance and negative perception regarding cooperatives have been found among members. For example, in a study in the late 1980's in parts of the present Limpopo Province, members were asked who owned cooperatives and only 41% indicated that they belonged to its members, while 23% didn't know. The other replies identified the government, chiefs, white people, the community or the management committee as owners (Machethe, 1990).

It can be concluded that group action by smallholders is required to obtain a more level playing field in the market. Stakeholders able to play a role in this respect are farmer cooperatives, value chain members, sector associations and NGOs.

Lack of institutional support

Each of the six case studies demonstrates the importance for smallholders of experiencing a strong institutional environment. Market imperfections, meaning that markets do not function properly according to principles of perfection, workability or contestable competition (e.g. Baumol, 1988; Christie, 2001), can be due to the lack of market institutions, as is the case with asymmetric market information among buyers and sellers or interlocked markets where farmers are constrained by credit ties in their choice of market outlet. Smallholders tended to suffer from asymmetric market information during the negotiation process with traders. Traders and auctioneers were more informed about current market prices than smallholders but this has been changed considerably due to modern information technology (e.g. Coetzee *et al.*, 2005). Markets for smallholders' inputs and outputs may be remote, consequently transport costs outweigh the margin that farmers can make. The conclusion can be that smallholders need to operate in a strong and supportive institutional environment in order to be able to access markets and to obtain a livelihood from their farming operations. They need capital for investments in farm equipment or they need working capital for their operational expenses before selling their crops or animals. Lack of purchasing power to buy farm inputs, e.g. veterinary services, will result in reduced output both in terms of quantity and quality and will, eventually, mean reduced profits. Stakeholders able to play a role in this respect are value chain members, sector associations, trade unions, national, provincial and district authorities, together with supporting research institutes and NGOs.

2.3.2 Bottlenecks for smallholders' access at market level

At market level, two types of bottlenecks for market access have been prevalent: market entry barriers and the failure to make use of market opportunities.

In an international context, market entry barriers have been reduced by WTO arrangements and agreements resulting in more competitive export markets. To exploit new opportunities, these arrangements and agreements need to be included in supply chain or sector marketing plans. For example, relatively new markets or market niches such as those for fair trade, organic produce and the developing ethanol energy market may create new opportunities, for smallholder farmers as well (e.g. Van Schalkwyk, 2007). The case study on rooibos tea is a good example. Marketing design studies such as the study on unlocking opportunities in the Eastern Cape province (CIAMD and ARC, 2001) can be used to broaden the access of small-scale farmers to agricultural markets. A possible conclusion could be that smallholders and their organisations need to be aware, or if necessary made aware, of opportunities in the market. Stakeholder groups able to be supportive in this respect are farmer co-operatives, value chain members, sector associations, national authorities, research institutes and NGOs.

2.3.3. Bottlenecks for market access at supply chain level

Market access through vertical coordination in the supply chain can be stimulated by the agribusiness or processing sector, by the retail sector (e.g. supermarket chains), or by export-oriented supply chains.

Market access through contract arrangements with the agribusiness sector

Purchasing agribusiness companies are interested in buying from suppliers who are able to meet their procurement requirements. A study on the potential for contracting arrangements of emerging farmers (Vermeulen *et al.*, 2007), based on interviews with 61 agribusiness companies in South Africa, concluded that contracting is an important instrument for emerging farmers to access supply chains. Almost 80% of the total volume of fruit and vegetables procured by agribusiness companies for processing, e.g. canning, drying, freezing, juice extraction, jam, sauces and snack foods, was based on some form of contractual arrangement. The balance was procured through a combination of open market transactions, their own estates or imports. South African retailers sourced 70 to 100% of their fresh produce directly from farmers. Meat, poultry and eggs were mainly procured through vertical integration, contracts and long-term informal supply arrangements with selected groups of farmers. Large abattoirs procured pigs through a combination of contract arrangements and spot market buying. Contracted farmers supplied up to 50% of the chickens to the major broiler processing companies. About 25% of the total volume of eggs procured by the major egg companies was based on contracting.

Market access to supermarket chains

The expansion of supermarkets in Kenya (Neven *et al.*, 2005) led to the development of a new group of medium-sized farms, managed by well-educated farmers. Nearly all farmers delivering to the supermarket channel had the capacity to supply larger volumes all year round. They had access to irrigation, transportation vehicles, a packing shed and a mobile phone, which points to a threshold capital vector needed by farmers before obtaining access to the supermarket channel. While most farmers in the traditional channel sold to brokers and received a price that allowed them at best to break even, farmers delivering to the supermarket channel had considerably higher gross profit margins. This resulted in strong growth for farmers delivering to the supermarket channel who doubled the size of their operations within five years.

> A major threat to micro-enterprises is the globalisation of the retail system and the rapid rise of supermarkets....which have moved to preferred supplier systems in which they select suppliers who are capable of meeting quality and safety standards (Bussard and Uhrinova, 2006: 56).

A problem that emerging farmers may encounter is that they experience competition from supermarket chains expanding into rural areas (e.g. Botha, 2006; Botha and Van Schalkwyk, 2007). Botha and Van Schalkwyk (2007) studied the potential market access of smallholders to the supermarket channel in Africa. The spread of supermarkets in Africa has been due to expanded services delivery. The introduction of dynamic supermarket chains can have both positive and negative effects on smallholders' market access. A positive influence is assumed when smallholders can deliver to supermarkets, thus providing local people with convenience shopping and access to low cost, good quality, products. A negative influence is assumed when supermarkets source their agricultural products elsewhere and compete with traditional local markets. The conclusion of the study is that both consumer welfare and economic growth benefited from the expansion of supermarkets into Africa (e.g. Reardon *et al.*, 2004). See also the Chapter 7 and Chapter 9 in this book for further illustrations.

Access to export markets

With the emergence of food safety and quality standards in developed countries, those countries exporting fresh food face increasing constraints to access markets in the North (e.g. Ruben *et al.*, 2007; Van Tilburg *et al.*, 2007; Vermeulen *et al.*, 2006). Producers in the South need to comply with the required quality and safety standards by making the necessary investments both on farms and in packing stations. Export of fresh fruit is a main component of South African agriculture. Vermeulen *et al.* (2006) studied the impact of compliance to private standards on both the quality of fruit and returns to farmers. Sampled fruit containers were followed throughout the supply chain to review the behaviour of the actors in the citrus supply chain and to obtain evidence on the handling and hygiene standards utilised. Observations suggested that the standards were adequately applied to the production and handling of fruit at both farm and pack house levels, but that subsequent stages of the fruit supply chain, mainly after the importing harbour in Europe, did not adhere to the same strict requirements laid out for producers.

The case studies on sugarcane, agricultural services and wool indicate the importance of a supply relationship between the smallholders and the processing industry. The rooibos tea case is a good example of the importance of an export market for smallholder farmers. The conclusion is that there are opportunities for smallholders to obtain access to agribusiness companies through contracts, provided that they manage to meet the companies' procurement requirements in terms of quality, quantity and delivery discipline. It can also be concluded that when a supermarket is established in their area, smallholders suffer if they do not succeed in becoming a supplier for this supermarket. Another conclusion can be that both food quality and safety standards set by import markets need to be applied equally at all points of the supply chain. Stakeholders who can facilitate access to supply chains are notably value chain members at different levels, sector organisations, public authorities and NGOs.

2.4 Stakeholder strategies to improve smallholders' market access

The focus of this section is to sketch potential stakeholder strategies to improve smallholders' market access. It follows the line taken in Section 2.3 by taking a bottleneck as the point of departure and discusses the roles the stakeholders or groups can or should play to become change agents. Next, the discussion is focussed on a suggested strategy to reduce or eliminate the influence of the bottleneck and, where possible, strategies and actions to be taken are discussed.

The following categories of private stakeholders may be able to deliver services to facilitate smallholders' market access: farmer cooperatives; value chain members, e.g. input suppliers, agribusiness processing companies, export companies or retail chains; sector associations, e.g. the Maize Association, Poultry Association, SA Grain; trade unions, e.g. the National African Farmers Union (NAFU) and non-governmental organisations (NGOs). The following categories of public stakeholders may be supportive to deliver services to facilitate smallholders' market access: national, provincial or district authorities, e.g. the provincial Departments of Agriculture; research institutes and universities or colleges involved in research to benefit small-scale or emerging farmers.

2.4.1 Smallholders access to resources: human, social and economic capital

Opportunities and stakeholder groups

Rural development initiatives that involve the beneficiaries in identifying their own alternatives and supporting them in achieving these through human and market development tend to succeed, unlike projects where community members are asked to participate but where the project is developed outside the community.

Four out of six case studies demonstrated the importance of improved access of smallholders to human, social and economic resources. There are opportunities for smallholders to deliver produce to interested parties provided that they meet the procurement requirements in terms of quality, quantity and delivery discipline. Farmer cooperatives, value chain members, national, provincial or district authorities and non NGO's can play a role. In this respect, there are several examples of smallholders supporting institutions in preceding sections:

- The joint action training and supporting model in the case study on the Suid Bokkeveld tea growers, and mentorship programmes such as the development of irrigation schemes by Provincial Departments of Agriculture.
- The agribusiness training and supporting model in the Temo Agri Services approach.
- The industry sector training and supporting model, for example, with respect to the case studies on sugarcane growers in KwaZulu-Natal, and the sheep farmers on communal land in the former Transkei.

Strategy and actions to be taken

The private sector needs to develop a plan for each sector, or to update an existing plan to empower smallholders in developing their human skills and farm management tasks with respect to knowledge about market opportunities, agricultural practices, production and financial planning.

2.4.2 Lack of group action

Opportunities and stakeholder groups

Each of the case studies illustrated the importance of group action by smallholders to obtain market access, for example, to obtain a more level playing field in the market and to reap the economies of scale. Stakeholders who can play a role in this respect are farmer cooperatives, value chain members, sector associations and NGOs.

Strategy and actions to be taken

As soon as smallholders become aware of the need for group action (e.g. in the case of the Heiveld Cooperative) they need to be organised to strengthen their countervailing power in the supply chain or with regard to spot market deliveries. Action that can be taken, mainly by the private sector, is to provide training and other support, in various models of group action, with the aim that each farmer group will be able to make the optimal choice in a particular situation.

2.4.3 Lack of institutional support

Opportunities and stakeholder groups

Each of the six case studies demonstrated the importance of a strong institutional context for smallholders to obtain a livelihood from their farming operations. Potentially successful farmers need to be embedded in an institutional context where a complete package of supporting services can be obtained at competitive prices.

Strategy and actions to be taken

Obtaining both sufficient purchasing power and access to farm inputs (e.g. veterinary services) in order to meet market requirements, it is essential to be able to run a viable business. Smallholders, being entrepreneurs, need both capital for investments and working capital to bridge the gap in time between sales and payments for operational inputs. Stakeholders that can play a role in this respect are the input supply industry, formal or informal finance institutions, sector associations, public authorities and NGOs.

2.4.4 Access to markets

Opportunities and stakeholder groups

Smallholders and their organisations need to become aware of opportunities in the market. We make a distinction between two product classes which require different strategies for market access: *commodities* such as cereals, soybeans and meat, through spot markets or through a processing agribusiness industry; and *niche market products* such as flowers, medicinal plants or herbal tea by entering a dedicated supply chain.

Strategy and actions to be taken

Two principal strategies with respect to the volume and value dimension are distinguished in obtaining market access. *Strategy A,* which is in line with Porter's low cost strategy in which smallholders with *high volume – low value* crops such as maize, wheat and bulk fruits are linked to commodity markets or supply chains through commercial co-ops. *Strategy B,* which is in line with Porter's differentiation strategy in which smallholders with *low volume – high value crops* are linked to niche markets through integrated supply chains. Examples of strategy A were represented by the case studies on sugarcane and wool, which demonstrated the importance of a supply relationship between smallholders and the processing industry. The rooibos tea case is a good example of strategy B.

Reviewing opportunities in South Africa with respect to strategy B identified the following *low volume – high value* products as opportunities for niche markets: herbal tea, e.g. rooibos tea, honeybush tea and greenbosch tea; organic herbs (e.g. dill, basil, sage and thyme); spices (e.g. cinnamon, coriander and cloves); medicinal plants (e.g. *hoodia gordonii*, an appetite suppressant); mushrooms; flowers or flower bulbs (e.g. lacenalia or *hippeastrum* or amaryllis); special seeds (e.g. pumpkin seed or sesame seeds, including their extracted oils); raisins; honey; handicrafts (e.g. Dawson, 2003); snails and mussels.

Actions to be taken for both strategies include:

- Knowledge transfer through vocational training (workshops) for farmers, covering issues related to extension, business development services and mentorship programmes involving experienced commercial farmers.
- Incentives for organisations to provide services to smallholders such as an award of the year for best example of emerging farmers support, or specific subsidies.
- Facilitating access to financial institutions such as commercial banks, the Land Bank, the Development Bank of South Africa, or the Industrial Development Cooperation.
- Collection and dissemination of market knowledge such as national and/or international market studies by sector organisations; or studies on the potential procurement of niche markets in or outside South Africa, e.g. on the market potential for medicinal applications of indigenous plants; allocation of research activities to support smallholders in their personal development, their management skills and their market knowledge by institutes, universities or colleges; and strengthening institutions dealing with rules and regulations affecting the smallholder's environment, for example: the Agricultural Strategic Plan, the Land Reform and Agricultural Development (LRAD) activities (e.g. Jordaan and Jooste, 2003), equal opportunity initiatives such as the AgriBEE rules and regulations, investment plans such as the Accelerated and Shared Growth Initiative in South African Agriculture (ASGISA), and access to investments funds through finance and micro-finance initiatives such as MAFISA.

2.5 Conclusions

The aim of the chapter has been to discuss and find strategies to deliver on market access for smallholders by discussing the following research questions: What needs to be done to improve market access for smallholder produce? Which stakeholders can contribute to facilitate smallholder access to markets or supply chains? What can be the role of each stakeholder group? Which promising alternative strategies can be developed to deliver on market access? As a result, the following conclusions and recommendations have been generated:

- There are clear examples or models in South Africa's agriculture to show how smallholders can be linked to markets in a sustainable way.
- Two promising strategic options, the *high volume – low value* strategy for commodities and the *low volume – high value* strategy for niche products need to be further developed and implemented in each sector in close cooperation with specific stakeholder groups.
- To be successful, intensified support is required for smallholder groups by means of both private and public initiatives and investments.

- Financial institutions (e.g. development banks) should further facilitate the initiation of research and implementation of promising projects, which would benefit smallholder market access.
- Sector organisations, agribusiness enterprises and member-cooperatives should intensify their initiatives to start projects for smallholders or emerging farmers in their sector.
- A multi-institutional task force for each sector, including representatives of emerging and commercial farmers, researchers, sector organisations and public authorities, needs to monitor the process of improving smallholders market access. This can be done in a way similar to that suggested in the Strategic Plan for South African Agriculture 2001 where goal orientation, sound planning, proper coordination, capacity building and monitoring of progress are mentioned as important tasks.

The analysis in, and conclusions of, this chapter can be used to plan a process of scaling up (e.g. Van Tilburg *et al.*, 2011) of promising experiments with respect to smallholder market access in South Africa.

References

Adey, S., 2007. A journey without maps: towards sustainable subsistence agriculture in South Africa. PhD thesis Wageningen University, Wageningen, the Netherlands.

Bates, R. and P. Sokhela, 2003. The development of small-scale sugarcane growers: a success story? In: L. Nieuwoudt and J. Groenewald (eds.) The challenge of change: agriculture, land and the South African economy. University of Natal Press, Pietermaritzburg, South Africa, pp. 105-118.

Baumol, W.J., J.C. Panzar and R.D. Willig, 1988. Contestable markets and the theory of industry structure, revised edition. HBJ publishers, Orlando, FL, USA.

Botha, L., 2006. The evolving food retail industry and the buying behaviour of consumers in developing areas: a Qwaqwa case study. MSc thesis in agricultural economics, University of the Free State, Bloemfontein. South Africa.

Botha, L. and H.D. Van Schalkwyk, 2007. An inquiry into the evolving supermarket industry in Africa. Paper presented at the International Food and Agribusiness Management Association's 17th Annual World Symposium, Parma, Italy, June 23-24, 2007.

Bussard, A. and L. Uhrinova, 2006. The problem of market access for MSEs in transition and developing countries. Small Enterprise Development 17: 52-61.

Christie, R.D., 2001. Evaluating the economic performance of alternative market institutions: Implications for the smallholder sector in Southern Africa. Agrekon 40: 522-536.

Centre for International Agricultural. Marketing and Development/Agricultural Research Council (CIAMD and ARC), 2001. Unlocking opportunites in the Eastern Cape Province. Unpublished report, University of the Free State, Bloemfontein, South Africa.

Coetzee, L., B.D. Montshwe and A. Jooste, 2005. The marketing of livestock on communal lands in the Eastern Cape province: constraints, challenges, and implications for the extension services. South African Journal for Agricultural Extension 34: 81-103.

Darroch, M.A.G. and M.C. Mashatola, 2003. Sugarcane growers' perceptions of a graduated mortgage loan repayment scheme to buy farmland in KwaZulu-Natatl, South Africa. International Food and Agribusiness Management Review 145: 353-365.

Dawson, J., 2003. Facilitating small producers' access to high-value markets: lessons from four development projects. Small Enterprise Development 14: 13-25.

Department of Agriculture, 2001. The strategic plan for South African agriculture. Government of South Africa, Department of Agriculture, Pretoria, South Africa.

D'Haese, M., W. Verbeke, G. Van Huylenbroeck, J. Korsten and L. D'Haese, 2003. Institutional innovations to increase farmers' revenue: a case study of small-scale farming in sheep, Transkei region, South Africa. IAAE Conference, Durban, South Africa.

D'Haese, M., G. Van Huylenbroeck, O.T. Doyer and M. Calu, 2007. A netchain development perspective on woolfarmers' associations in poor communities: a case study in South Africa. Journal on Chain and Network Science 7: 11-20.

Femi, O.A. and H.D. Van Schalkwyk, 2006. Mentorship alliance between South African farmers: implications for sustainable agriculture sector reform. IAAE Conference, Gold Coast, Australia, 13 p.

Henderson, J.M. and R.E. Quandt, 1980. Micro-economic theory; a mathematical approach, 3rd ed. McGraw-Hill, New York, NY, USA.

Hill, B.E. and K.A. Ingersent, 1982. An economic analysis of agriculture, 2nd ed. Heinemann Educational Books, London, UK.

Holloway G, C. Nicholson, C. Delgado, S. Staal and S. Ehui, 2000. Agroindustrialization through institutional innovation: transaction costs, cooperatives and milk-market development in the east-African highlands. Agricultural Economics 23: 279-288.

Jordaan, A.J. and A. Jooste, 2003. Strategies for the support of successful land reform. A case study of Qwaqwa emerging commercial farmers. South African Journal of Agricultural Extension 33: 1-14.

Machethe, C.L., 1990. Factors contributing to poor performance of agricultural co-operatives in less developed areas. Agrekon 29: 305-309.

Magingxa, L.L., 2006. Smallholder irrigators and the role of markets: a new institutional approach. PhD thesis, University of the Free State, Department. of Agricultural Economics, Bloemfontein, South Africa.

Magingxa, L.L., Z.G.Alemu and H.D. Van Schalkwyk, 2006. Factors influencing the success potential in smallholder irrigation projects of south africa: a principal component regression. International Association of Agricultural Economists Conference, Gold Coast, Australia, August, 2006.

Nel, E., T. Binns and D. Beck, 2007. 'Alternative foods' and community-based development: Rooibos tea production in South Africa's West coast mountains. Applied Geography 27: 112-129.

Neven, D., Th. Reardon, M. Odera and H. Wang, 2005. Farm-level perspectives on the impact of domestic supermarkets in Kenya's fresh fruit and vegetables supply system. MSU Staff Paper 2005-05, Michigan State University, East Lansing, MI, USA.

Nothard, B.W., G.F. Ortmann and E. Meyer, 2005. Institutional and resource constraints that inhibit contractor performance in the small-scale sugarcane industry in KwaZulu-Natal. South African Journal of Agricultural Extension 34: 55-80.

Ortmann, G.F and R.P. King, 2007a. Agricultural cooperatives I. Agrekon 46: 40-68.

Ortmann, G.F and R.P. King, 2007b. Agricultural cooperatives II. Agrekon 46: 219-244.

Reardon, T., C.P. Timmer, and J.A., Berdegue, 2004. The rapid rise of supermarkets in developing countries: induced organizational, institutional and technical change in agrifood systems. Journal of Agricultural and Development Economics 1: 168-183.

Ruben, R., M. Van Boekel, A. Van Tilburg and J. Trienekens, 2007. Tropical food chains: governance regimes for quality management. Wageningen Academic Publishers, Wageningen, the Netherlands.

Sartorius, K. and J. Kirsten, 2002. Can small-scale farmers be linked to agribusiness? The timber experience. Agrekon 41: 295-325.

Stern, L.W., A.I. El-Ansary and A.T. Coughlan, 1996. Marketing channels, 5th ed., Prentice Hall, Upper Saddle River, NJ, USA.

Swart, D., F.O. Hobson and J.P. Carstens, 2000. A best-practice model for agricultural and rural community development in the Eastern Cape province of South Africa: the Transkei and Ciskei example. Third All African Conference on Animal Agriculture, November 2000, Alexandria, Egypt, p. 5.

Tregurtha, N.L and N. Vink, 1999. Trust and suply chain relationships: a South African case study. Agrekon 38: 755-765.

Van Rooyen, C.J., 1984. The identification and attitudes of successful and less successful farmers in smallholder agriculture in Ciskei. Agrekon 23: 14-19.

Van Schalkwyk, H.D., 2007. A new vision for Land Bank. Land Bank of South Africa, Pretoria, South Africa.

Van Schalkwyk, H., J. Groenewald and A. Jooste, 2003. Agricultural marketing in South Africa. In: L. Nieuwoudt and J. Groenewald (eds.) The challenge of change: agriculture, land and the South African economy. University of Natal Press, Pietermaritzburg, South Africa, pp. 119-135.

Van Tilburg, A., J.H. Trienekens, R. Ruben and M.A.J.S. Van Boekel, 2007. Governance for quality management in tropical food chains. Journal on Chain and Network Science 7: 1-9.

Van Tilburg, A., 2010. Linkages between theory and practice of marketing in developing countries. In: H. Van Trijp and P. Ingenbleek (eds.) Markets, marketing and developing countries: where we stand and where we are heading. Wageningen Academic Publishers, Wageningen, the Netherlands, pp. 164-184.

Van Tilburg, A., E. Kambewa, A. De Jager and D. Onduru, 2011. Up-scaling smallholder participation in global value chains. In: A.H.J Helmsing and S. Vellema (eds.). Value chains, inclusion and endogenous development: contrasting theories and realities. Routledge, Abingdon, UK, pp. 247-265.

Vermeulen, H., D. Jordaan, L. Korsten and J. Kirsten, 2006. Private standards, handling and hygiene in fruit export supply chains. A preliminary evaluation of the economic impact of parallel standards. IAAE Conference, Gold Coast, Australia, 14 p.

Vermeulen, H., J. Kirsten and K. Sartorius, 2007. Contracting arrangements in agribusiness procurement practices in South Africa. Paper presented at the 45th annual conference of the AEASA, 25-27 September, Johannesburg, South Africa.

Appendix 2.1. Case study 1: a marketing design study

Summary

Marketing design studies are conducted with the aim of broadening the access of small-scale farmers to agricultural markets. An example of such a study is the CIAMD and ARC (2001) study on unlocking opportunities in the Eastern Cape province.

The aim of this study was to investigate the competitiveness of the agricultural sector of the Eastern Cape and to identify options and opportunities to broaden the agricultural base of the province. Specific objectives were to obtain a thorough understanding of the agricultural sector; to identify opportunities to alleviate poverty within the context of the resource base; to assess options in terms of their ability to be economically feasible, to generate new employment opportunities, to promote small enterprise development,to create links with other sectors in the provincial economy; to rank these options on criteria to generate employment, return on capital investment, access to input, service and output markets, and to develop good leadership and democratic membership for group action of small-scale farmers.

The CIAMD and ARC (2001) study examined potential outgrower schemes in which commodities are grown under prescribed production guidelines and predetermined marketing arrangements between producers and a buyer. It also considered the profitability of integrated livestock and crop production systems. It was considered that small-scale farmers need to be organised into legal entities to take full advantage of economies of scale. Commodities considered in this study included livestock, bulbs, flowers, herbs and medicinal plants. The following example illustrates the CIAMD-ARC approach.

The Roxeni village community (near Alice in the central region of the Eastern Cape) has been engaged in various agricultural activities aimed at alleviating poverty (CIAMD and ARC 2001, Chapter 4). The Roxeni community project initiative included five projects for sustainable rural community development regarding vegetables, sheep, beef, pigs and poultry. The community was well organised, members were dedicated under good leadership, human capacity was strengthened through training sessions, support services offered were utilised, and members appeared to be creative in exploring value-adding options.

The project started with community meetings to spell out the project objectives and to develop clauses on membership fees. The constitution drafted included the phases of project implementation to enhance efficiency and competitiveness, sourcing of capital and training, facilitation and development of infrastructure, the establishment of community based demonstration units, effective classing and marketing of wool, and consolidation and expansion of successes.

Bottlenecks and lessons learnt

The main bottlenecks for these smallholders were a lack of insight into market opportunities in relation to their own strengths and weaknesses, a lack of proper group action and shortcomings in the enabling environment. The main lessons learnt included: rural development projects that involve beneficiaries in identifying their own choice of alternatives tend to succeed, unlike projects where the project is developed outside the community and the community members are asked to participate. Another lesson learnt from this study was that unlocking Eastern Cape provinces' potential for smallholders or emerging farmers in particular was highly dependent on creating an enabling environment.

Appendix 2.2. Case study 2: sugarcane production by smallholders in KwaZulu-Natal

Summary

The sugar industry in KwaZulu-Natal, presents an example of how the agribusiness sector and small-scale farmers can benefit from promoting the development of these small-scale sugarcane growers by the sugar industry. The case illustrates the importance of a well-developed institutional environment for smallholder development (Bates and Sokhela, 2003; Nothard *et al.*, 2005).

Smallholders

The South African government initiated in 1956 an assistance programme for small-scale sugarcane growers (SSGs) by providing finance for ploughing and the purchase of inputs. Sugarcane quotas were issued In the period 1963-1967 to SSGs allowing them to plant about 7,000 ha. The sugar industry introduced an expansion programme for the development of the small-scale grower sector from the 1970s onwards, and the number of SSGs increased from about 4,000 in 1972-1973 to about 50,000 in 1994-1995. A credit programme was introduced in 2001 providing additional production incentives. During the 2001-2002 season, SSGs producing sugarcane in communal areas comprised 96% of the growers in the sugar industry, they produced about 14% of the sugarcane crop on 20% of registered sugarcane land. Up to 2002, several sugar mills provided contractual ploughing and other management services. However, the withdrawal of sugar mill contracting services contributed to a decline in SSG production.

Contractors

Small-scale sugarcane contractors, usually being SSGs themselves, diversified their activities by providing land preparation, crop maintenance and cane haulage services as well as labour contracting services to SSGs Several inhibiting factors of contractor performance were found in a survey among 124 small-scale sugarcane contractors (2002-2003) such as high transaction costs in sourcing and operating contract work; limited access to finance without collateral; cash flow problems due to delays in payments; asymmetric bargaining power by contractors (men) with SSGs (often women); limited investments in new machinery and tractors; lack of labour for cane cutting and a poor rural infrastructure. Recommendations for a more competitive contractor sector included allowing contractors to source their own work and making SSGs aware that they can demand high quality services as customers of the contractors.

The sugarcane industry

The periodical *Sugarcane growers* (2007) reported several initiatives aimed at supporting small sugarcane growers: preferential product prices through the small-scale grower Supplementary Payment Fund; fair market access to available milling capacity through legislation (Sugar Act and Sugar Industry Agreement); training, extension and support through the Shukela Training Centre; contractor support programmes to improve the capacity of contractors in servicing the small-scale growers; Umthombo Agricultural Finance to make savings and loans services accessible; improved rural infrastructure (loading zones, roads, bridges, communications) to facilitate contractor access and harvesting; improved access to farm inputs through decentralisation of agricultural cooperatives and industry grant funding to establish seed cane nurseries.

Another development for individual small-scale growers (*Sugarcane growers*, 2007) has been the consolidation of their arable land under a cooperative to generate economies of scale. In this programme, each individual grower leases his land to the Cooperative, which runs the consolidated area according to commercial agricultural principles and, in return, the grower receives an annual rental for his land and/or dividend payment based on profitability. These cooperatives appeared to be effective vehicles to attract funding. About 16% of freehold land under sugarcane has been transferred to black growers (*The South African Sugar Industry Directory 2007/08*)[2]. The industry established a land reform entity to support the transfer of ownership by identifying both sellers and buyers. The industry supported SSGs operating on tribal land with mentorship and training programmes on technical skills to counteractproblems with planting, growing, cutting and cane delivery to the mills.

Bottlenecks and lessons learnt

The main bottlenecks were a lack of supporting services to sugarcane smallholders, limited investments by small-scale sugarcane contractors, lack of contractor market transparency and asymmetric bargaining power between contractors and smallholders. Lessons learnt

[2] Available at: www.sugar.org.za/subscribe/downloads/Sugarindustry2007.pdf.

identified the need to strengthen farm management abilities of SSGs through training and mentorship, the need for investments in machinery by contractors and the need to improve contractor market performance by improved market transparency and more symmetric bargaining power between SSGs and contractors.

Appendix 2.3. Case study 3: TEMO Agri Services

Summary

The goal of this initiative taken by the agribusiness company MGK in Britt has been to assist emerging farmers on their way to become commercial farmers.

MGK, an agribusiness company founded in 1930, established Temo Agri Investments that obtained 22% of the MGK shares in a Black Economic Empowerment transaction. One of Temo Agri Investments shareholders is a Farmers Trust, established for the benefit of emerging farmers. Temo Agri Services (TAS)[3] is a joint venture of both MGK and the Temo Farmers Trust. TAS aims to 'bring the farmer back to the land' by helping land owners to use their land for farming purposes and it gives the MGK company a new client base. The role of Temo Agri Services is to select potential successful emerging farmers, to provide these farmers with *mentor services* (training) and to try to also secure other support (including financial support) for these participating farmers. The main criteria for farmer selection are: motivation, land availability, farmer's focus on the farming business, and the land available must be suitable for growing grain and oilseeds. All decisions on land preparation, crop selection and cultivars, planting and harvesting, hiring of contractors, *are taken jointly* by farmer and mentor farmer. All participating emerging farmers are organised into study groups that have meetings to solve problems, training and planning sessions on both farming processes and communication, with services to be provided by MGK divisions. A supporting TEMO-MGK division can provide the required services to member emerging farmers, e.g. finance (credit), crop insurance, necessary inputs and marketing services. All participating farmers have a compulsory crop insurance against risks related to hail and fire. All necessary farm inputs, the hiring of both accredited contractors and extension specialists can be obtained on credit. The MGK marketing division supports emerging farmers by providing price information and hedging of prices through Safex, transport of crops from farm to silo, and silo storage of crops until final marketing.

[3] Company website: http://www.temoagri.co.za.

Bottlenecks and lessons learnt

The main bottlenecks for smallholders were access to and the utilisation of land, farm management abilities, lack of group action, the provision of a complete package of inputs and market access.

The main lesson learnt was, that potential emerging farmers need to be trained in farm management practices and be embedded in a proper institutional context, in order to qualify for a complete package of supporting services.

Appendix 2.4. Case study 4: linking emerging farmers to the agribusiness sector – the case of the Golden Fleece project in the wool industry

Summary

Commercial producers, brokers, exporters and spinners dominated the wool supply chain in South Africa. Smallholder farmers in the Transkei region had limited access to a profitable market outlet for their wool. In response, the South African wool industry took the initiative to support local farmers by building shearing sheds under which the local association can bulk the wool and trade directly with brokers. More direct access to the wool brokers appeared to be a prerequisite for these farmers to develop a viable business (D'Haese *et al.*, 2003, 2007; Swart *et al.*, 2000; Van Rooyen, 1984).

The wool marketing initiative was taken by the National Wool Growers Association (NWGA)[4], a national commodity structure of commercial and communal wool sheep farmers in South Africa representing more than 80% of national wool production. Close to 50% of the members are black communal or emerging farmers. The aim of the project was to improve both quantity and quality of wool production together with increased market efficiency for both the Transkei and Ciskei farmers and the National Wool Industry. Thirty-two shearing sheds were built, which also served as training and development centres. They provided animal handling facilities, dipping tanks and other critical agricultural services. During 1998 and 1999 more than six thousand sheep farmers received practical training in farming technology, wool production practices, product value adding and marketing. Successful aspects of the programme were increased weaning percentages and a ram-breeding project. Household income from wool production tripled and an approximate increase of 30% in available jobs in wool production and handling were noticed as direct results of the project. The new infrastructure and services markedly increased the production and income generation potential of the Transkei and Ciskei communities.

[4] NWGA website: http://www.nwga.co.za.

Bottlenecks and lessons learnt

The main bottlenecks were lack of skills to add value, lack of market access, lack of group action and lack of economies of scale. The main lessons learnt were that improved market access could be obtained through industry initiatives to promote group action and the building of shearing sheds, which also serve as training centres.

Appendix 2.5. Case study 5: emerging rooibos farmers in the Heiveld community on the South Bokkeveld plateau

Summary

Two marginalised communities in the Northern Cape province, the Heiveld community on the South Bokkeveld plateau, east of the Cedarberg, and the Wuppertal community in the Cedarberg mountains have successfully penetrated international markets by supplying organically produced Rooibos tea which is certified by the international Fair Trade System (Adey, 2007; Nel *et al.*, 2007).

Until the early 1990s, South Africa's agricultural production was managed by agricultural boards, which generally favoured white producers and marginalised other farmers. In 1991, the industry was deregulated, allowing new commercial producers to enter the market. Following democratic elections in 1994, the marketing of tea remained largely monopolised by the Rooibos Tea Cooperative. The farmers in the above two communities could only deliver to commercial farmers when the cooperatives could not meet their own quota. The market breakthrough for small-scale community producers came when they were targeted for support by two NGOs, which played critical roles in facilitating the two communities in producing a commodity that is environmentally sustainable, meets ethical criteria and is produced for the international market. One of these two cooperatives is the Heiveld Cooperative. Following the establishment of the Heiveld cooperative, the Suid Bokkeveld smallholders adopted the same ideas and, following meetings with the NGOs and appropriate training, the Suid Bokkeveld smallholders formed their own cooperative.

The steps leading to the ultimate registration of the Heiveld Cooperative Ltd. were the initial community meeting (March 1998), knowledge-exchange visits, the decision to form a collective organisation (August 2000), a planning workshop and the election of a management committee (September 2000), interest of Fair Trade in rooibos tea (November 2000), a three year contract with a local tea trader and the first export of fair trade organic rooibos tea (December 2000), the formation of the Heiveld Small Farmers Association (January 2001), the registration of the Heiveld Cooperative Ltd, and the direct export of rooibos tea to European alternative trade organisations. The required knowledge and skills obtained by the Suid Bokkeveld farmers to become commercial farmers were partly based on local knowledge of organic rooibos cultivation by people working on their own land

and those working as labourers on commercial farms. Knowledge of the organic cultivation of rooibos, post-harvest treatment and marketing were integrated in a social learning process fostered by NGOs and researchers. This was facilitated by the fact that the project, initiated in post-apartheid South Africa, provided a suitable environment for participatory approaches and for mobilising resources to address the needs of emerging farmers.

Bottlenecks and lessons learnt

The main bottlenecks were market access, knowledge about how to produce and deliver tea according to the requirements of an international buyer, and lack of group action to strengthen both local knowledge and social organisation leading to smallholders' economic development. The lessons learnt were that NGOs played critical roles in facilitating the two projects: social capital has been strengthened; the communities have gained in confidence and enhanced their production, marketing and management skills. NGOs also played a crucial role in linking poor communities to sources of external funding and potential markets.

Appendix 2.6. Case study 6: mentorship alliance between South African farmers

Summary

Mentorship of black emerging farmers by white commercial famers can be a way to the success of land reform by reducing the knowledge gap between the more and less experienced farmer groups. Questions raised have been whether the objectives of a mentorship programme have been clear enough. What have been the required characteristics of mentorship programmes to be successful? And, what can or should be the compensation or reward for farmers delivering mentorship support to emerging farmers? (e.g. Darroch and Mashatola, 2003; Femi and Van Schalkwyk, 2006). The papers discuss several mentorship programmes in the Free State province and KwaZulu-Natal. The conclusion is that forming interactive mentorship alliances that are complementary, loosely structured, based on previous experience of smallholders, and not being hindered by complicated rules and regulations can be successful, provided that they operate in an enabling environment with opportunities for group action by emerging farmers and a level playing field in markets relevant for these farmers.

Bottleneck and lessons learnt

A main bottleneck is the lack of training and skills among emerging farmers in coping with challenging production and marketing environments. The lesson learnt is that mentorship alliances can reduce the knowledge gap between commercial and small-scale farmers.

3. Influence of institutional and technical factors on market choices of smallholder farmers in the Kat River Valley

Bridget Jari and Gavin Fraser

3.1 Introduction

In the age of trade liberalisation and globalisation, the world markets are increasingly being integrated. This implies that farmers in the developing world are ever more linked to consumers and corporations of the rich nations. Consequently, local farmers are facing increasing market competition, not only in international markets but in local markets as well. In an effort to withstand the market pressures, agricultural markets are now transforming to a vertically coordinated structure (Reardon and Barrett, 2000). In addition, both the private and the public sectors have made some adjustments in agricultural markets, in order to survive competition resulting from market changes.

The South African agricultural sector deregulated in 1997, with the aim of creating an open and market-oriented environment in order to boost the sector. Based on the Agricultural Products Act of 1996, government intervention in agricultural marketing using control boards was ceased. This change resulted in smallholder farmers and other formerly deprived farmers in output markets being potentially included in agricultural marketing (Meyer *et al.*, 1998). Although policies are now oriented in favour of smallholder farmers, they still have to compete for markets with the already well-developed commercial sector. For this reason, their survival in the markets is still at stake. In output markets, smallholder and emerging farmers often face difficulties in enforcing contracts and meeting stringent food safety norms; they lack skills, are located in remote areas, and mostly rely on middlemen. They are usually served by poor physical infrastructure and weak institutions in markets (Kherallah and Kirsten, 2001; Makhura, 2001). Understanding such challenges among smallholder farmers is important in identifying areas that need focus and direction for improvement. In the light of these challenges, suggestions can be made on how to improve emerging and smallholder farmers' participation in output markets.

The main objective of this chapter is to identify and assess the technical and institutional challenges that are faced by smallholder agriculture in globalised output markets for farmers in the Kat River Valley. The study focuses on the factors that compel smallholder and emerging farmers to make certain marketing decisions. In other words, it considers factors that guide farmers in deciding whether to sell produce or not. It further looks at the factors that influence the choice of marketing channels when selling produce.

3.2 Importance of smallholder farmers

The importance of smallholder agriculture in developing countries is being recognised, which is the reason why there are countries where land is transferred to smallholders and development programs are redirected towards empowering these farmers (Dorward and Kydd, 2006; Lewis, 1954 cited in Todaro, 1997; Mellor, 1966 cited in Todaro, 1997). The proponents of smallholder farming argue that with enhanced market access, smallholder agriculture has potential to commercialise and contribute towards food security and poverty alleviation through food price reduction and employment creation. In addition, efficient smallholder agriculture leads to increased incomes and promotes equitable distribution of income, creates backward and forward linkages necessary for economic growth (Dorosh and Haggblade, 2003; Magingxa and Kamara 2003; Poulton *et al.*, 1998; Reardon and Barrett, 2000). In this way, the smallholder agriculture sector is not only important for the revitalisation of the agricultural sector, but for the economy at large.

In South Africa, the potential contribution of smallholder farmers to the economic growth remains locked. The major challenges facing smallholder agricultural growth are closely associated with lack of marketing knowledge and opportunities, calling for market-oriented interventions. In marketing, the smallholder agricultural sector still resembles past Apartheid legacy, where the sector has difficulties in marketing produce through formal channels (Carter and May, 1999).

3.2.1 Smallholder farmers in markets

Markets are important because they act as a mechanism for exchange, derive benefits such as income and open opportunities for rural employment (Dorward *et al.*, 2003; Machethe, 2004). Marketing activities such as processing, transportation and selling can provide employment for those willing to exit the farming sector. At national level, Lyster (1990) identified that the involvement of smallholder farmers in markets contributes to poverty alleviation and is important for sustainable agriculture and economic growth. Poverty is reduced through food price reduction, employment creation and farm income generation. Farmers' abilities to plough back their farm profits into the farm business result in sustainable agriculture and economic growth (Dorward *et al.*, 2003).

3.2.2 Overview of smallholder marketing channels

For farmers, growing and harvesting a crop and rearing animals form only half of the battle because they still have to market the produce. For smallholder farmers in South Africa, marketing produce remains a challenge. This group of farmers faces difficulties in marketing, even though individual smallholder farmers may be integrated with national or international markets (Shiferaw *et al.*, 2006). Before choosing a marketing channel, smallholder farmers consider the costs associated with transportation, profits, level of trust among the available brokers and familiarity of the markets, among other factors (Makhura, 2001). In other instances, farmers market their produce through channels offering low prices because they either lack market knowledge or have difficulties in accessing markets that are more rewarding.

Most produce from smallholder farmers in South Africa is sold locally, with only a small amount exported. Generally, smallholder farmers market their produce individually in local markets, but make use of market intermediaries in international markets. The marketing channels that are usually followed by smallholder farmers are illustrated in Figure 3.1.

The arrows in Figure 3.1 illustrate the different paths that are followed by produce harvested and sold by smallholder farmers until they reach the consumers. Produce from smallholder farmers is sold to consumers and traders at the farm gate, usually through informal transactions where prices and terms of exchange are unofficially negotiated. These transactions between farmers and traders and between farmers and consumers most often occur in spot markets (Kherallah and Minot, 2001; Ruijs, 2002).

When compared to vertical coordination in the supply chain, some weaknesses are associated with spot markets. For instance, prices and conditions of delivery are negotiated for every transaction carried out on spot markets. This may result in increased marketing costs for the farmer. Moreover, farm gate sales tend to result in lower farmer revenue since the prices are relatively low and variable (Montshwe, 2006). Variable prices may result from unavailability of scales for weighing produce, asymmetric market price knowledge and opportunistic behaviour by the more informed traders. In addition, at the farm gate, farmers may sell to

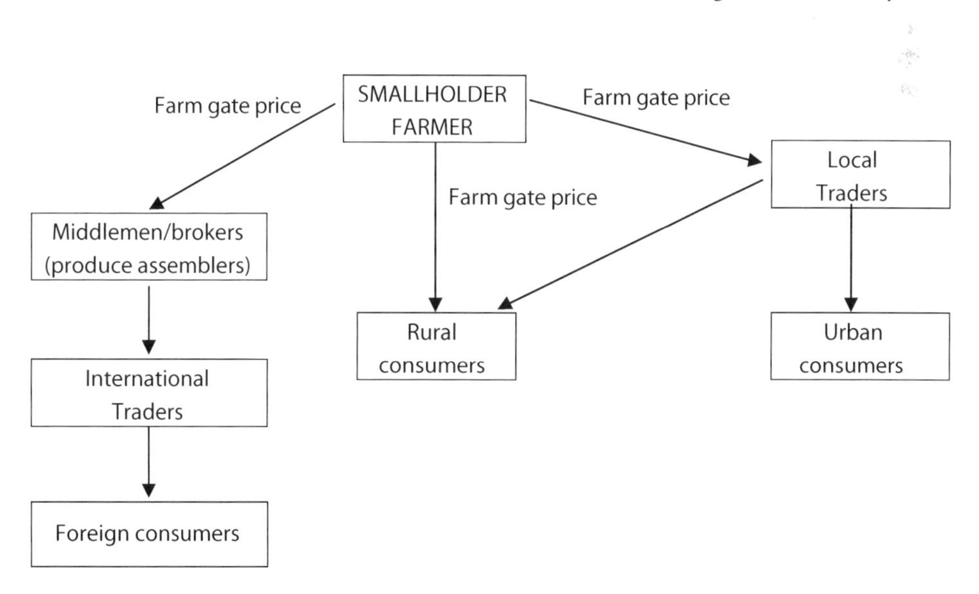

Figure 3.1. Marketing channels for produce from smallholder farmers (Simplified from Shiferaw et al.*, 2006).*

their neighbours, even when the latter cannot pay immediately for the produce. However, smallholder farmers tend to prefer farm gate sales because they receive immediate payments and do not incur marketing costs such as transportation costs and tax payments (Shiferaw *et al.*, 2006).

3.2.3 Challenges faced by smallholder farmers in output markets

Smallholder farmers face difficulties in accessing markets, and as a result, markets do not serve the interests of smallholder and emerging farmers. In South African's less developed rural areas, smallholder and emerging farmers find it difficult to participate in commercial markets due to a range of technical and institutional constraints. Factors such as poor infrastructure, lack of market transport, dearth of market information, insufficient expertise on, and use of grades and standards, inability to conclude contractual agreements and poor organisational support have led to inefficient use of markets, hence, results in commercialisation bottlenecks. Furthermore, smallholder farmers lack vertical linkages in the marketing channels, which result in their exclusion from the use of formal markets (Delgado, 1999; Fenwick and Lyne, 1999; Makhura, 2001; Wynne and Lyne, 2003). Smallholder farmers have weak financial and social capital and limited access to legal recourse, implying that it is difficult to change these negative market factors individually (Fenwick and Lyne, 1999). As a result, they are trapped and continue to operate within the given market constraints and they do not receive rewarding incomes from their agricultural activities.

Institutional aspects in smallholder agricultural markets

Institutions, defined in the framework of new institutional economics, are rules of the game, which have been formulated to govern relationships between individuals or groups of people involved in transactional activities (North, 1990). Institutions are divided into formal and informal institutions, where formal institutions refer to legal rules whereas informal institutions refer to non-legal rules that are enforced by peers. Market institutions are the underlying determinants of economic performance since they shape the organisation of market transactions (Kherallah and Kirsten, 2001).

Institutional aspects in marketing and economic development include transaction costs, market information flows and the institutional environment. Smallholder farmers in less developed rural economies lack adequate market information and contractual arrangements, lack lobbies in the legal environment and are not easily receptive to changes (Delgado, 1999; Kherallah and Kirsten, 2001). These factors tend to result in high transaction costs and, hence, difficulties in formal market participation. Where transaction costs are higher than the marketing benefits, smallholder farmers are discouraged from selling their produce using that particular marketing system (Makhura, 2001).

Transaction costs

Transaction costs are observable and non-observable costs associated with enforcing and transferring property rights from one person to another (Eggertson, 1990). These include the costs of searching for a trading partner with whom to exchange with, the costs of screening partners, of bargaining, monitoring, enforcement and, eventually, transferring the product to its destination (Hobbs, 1997; Jaffee and Morton, 1995). Delgado (1999) identified high transaction costs as the embodiment of market access barriers among resource poor smallholders. These high transaction costs result from individual produce transportation and selling, difficulties in getting trading partners and poor bargaining power (Delgado, 1999). When transaction costs are high, smallholder farmers may cease produce marketing. In other words, with high transaction costs, markets fail in their role of allocating scarce resources to alternative ends. For South Africa, Makhura (2001) explained that high transaction costs prevail among the smallholder farmers.

Market information

Market information is vital to market participation behaviour of smallholder farmers. Market information allows farmers to take informed marketing decisions that are related to supplying necessary goods, searching for potential buyers, negotiating, enforcing contracts and monitoring. Necessary information includes information on consumer preferences, quantity demanded, prices, produce quality, market requirements and opportunities (Ruijs, 2002). Of equal importance is the source of market information because it determines accuracy of the information.

According to Montshwe (2006), smallholder farmers have difficulties in accessing market information, exposing them to a marketing disadvantage. Smallholder farmers normally rely on informal networks (traders, friends and relatives) for market information due to weak public information systems (FAO, 2004). However, such individuals may not have up to date and reliable market information, making the usefulness of the information doubtful. Additionally, farmers relying on informal networks for market information are at risk of getting biased information due to opportunistic behaviour of the more informed group. For instance, Mangisoni (2006) explained that smallholders usually accept low prices for their crops when the broker informs them that their produce is of poor quality. Smallholder farmers accept these low prices mainly because they are unable to negotiate from a well-informed position.

Grades and standards

Consumers demand high quality for the goods they buy. In addition, they will not buy food products unless there is a guarantee that they are safe to eat (Kherallah and Kirsten, 2001).

In other words, consumers make purchasing decisions depending on packaging, consistency as well as uniformity of goods.

Most smallholder crops have no clearly defined grades and standards and, therefore, cannot meet the consumers' demands (Reardon and Barrett, 2000). Produce from smallholder farmers do not meet certain market grades and standards because the farmers lack the knowledge and resources to ascertain such requirements. In addition, institutions for determining market standards and grades tend to be poorly developed in smallholder farmers environments. Due to uncertainty on the reliability and quality of their goods, they usually cannot get contracts to supply formal intermediaries such as shops and processors (Benfica *et al.*, 2002). This indicates that only well organised farmers can benefit from trade liberalisation by adopting strict quality control measures and obtaining the necessary certification for their goods.

Organisation in markets

Smallholder farmers tend not to be organised in the markets as they usually sell their limited agricultural produce surpluses individually and directly to the consumers without linking to other market actors (Key and Runsten, 1999). In other words, smallholder farmers lack collective action in markets. Individual marketing of small quantities of produce weakens the smallholder farmers' bargaining positions and often exposes them to price exploitation by traders. They also do not benefit from economies of scale (Kherallah and Minot, 2001).

In a globalised world, there is increasing vertical integration and alliance formation in the agricultural marketing channels and markets, in an effort to meet consumer needs. Such alliances include contract farming, cooperatives and farmer organisations. Agribusiness firms favour contracts with medium to large-scale farmers, such that individual smallholder farmers cannot be part of these contracting arrangements (Key and Runsten, 1999; Kherallah and Kirsten, 2001). Lack of facilitation in the formation of producers associations or other partnership arrangements makes it more difficult for smallholder producers to participate in formal markets. The greater the degree of organisation in the market, the smaller the transaction costs are likely to be and the easier it is to benefit from the exchange opportunity (Frank and Henderson, 1992). Unfortunately, lack of collective action among smallholder farmers denies them entry into formal market channels.

Legal environment

Legal institutions influence the activities performed on the market and the costs of exchange. Minot and Goletti (1997) affirm that the formal institutional development of a society has a considerable influence on transaction costs. Thus, if trade laws are transparent then agreements can be legally enforced, leading to information accessibility and lower costs.

In other words, effective legal institutions may improve the organisation of the marketing channels and decrease marketing costs.

In many developing countries, laws are not always executed and enforced correctly, bribery and cheating are often not penalised, courts are out of reach for the majority of the population, and market rules are often not transparent to the producers and traders (Ruijs, 2002). In addition, formal contract enforcement mechanisms are weak (Fafchamps, 1996). It is even worse for the smallholder farmers because they lack lobbies in the legal environment. As a result, rural trade prospers where trust has been developed based on repeated transactions or informal relationships (Randela, 2005). Thus, an unfavourable legal environment creates a significant barrier to entry into formal food trade and limits participation by smallholders in the modern marketing system.

Government policies related to smallholder farmers

Prior to 1997, the agricultural marketing policy in South Africa was characterised by government support and controls (Kherallah and Kirsten, 2001). The success stories of the commercial farmers who received support during that time show that the government policies and incentives are influential and important in mobilising the farmers. For instance, the commercial farmers who received support from government policies still enjoy the legacy of the policy because they gained access to markets, an important condition in the deregulated markets.

On the other hand, the emerging and smallholder farmers do not have access to the levels of state assistance and market share, which the government previously guaranteed to commercial farmers. This implies that the smallholder and emerging farmers have to gain and compete for a market share on their own. Since these farmers are still learning, they face difficulties in competing in well-developed markets. The situation calls for a need of the government to create an enabling market environment for the smallholder farmers through selective financial government support, reduction in anti-competitive behaviour and facilitating private sector and farmer organisations partnerships. In addition, the government can ensure that public facilities, such as information and infrastructural facilities are developed in areas where smallholder and emerging reside, for their benefit (Kherallah and Kirsten, 2001; Nel and Davies, 1999).

Technical aspects in smallholder agricultural markets

Technical changes in marketing can be viewed as those transformations that allow goods to be available on the market at lower costs and in a more diversified set of markets. Technical changes are usually influenced by factors in the organisation itself, public regulation and general advances in technology. In agricultural production and marketing, smallholder farmers tend to be lagging in the use of improved technology (Carrè and Drouot, 2002).

Machethe (2004) pointed out that most small producers in South Africa lack appropriate transportation facilities and road infrastructure, communication links and storage infrastructure resulting in high transaction costs. Further, smallholder farmers have limited ability to add value to their produce. Lack of such facilities usually constrains farmers' supply response to any incentives in both agricultural production and marketing (Dorward *et al.*, 2003). Sometimes transaction costs are too high for farmers and traders to get any meaningful benefits from potential trading activities, discouraging farmers to participate in marketing activities.

Storage facilities

The ability to deliver a quality product to the market and ultimately to the consumer, commands buyer attention and gives the grower a competitive edge (Bachmann and Earles, 2000). Proper post harvest handling and storage contribute in ensuring quality maintenance for perishable agricultural produce. Moreover, agricultural commodities have to be harvested at a specific point in time, but are consumed year-round, thus necessitating proper storage facilities (Sasseville, 1988). Therefore, if crops are to be available for consumption throughout the year, proper storage facilities have to be implemented by both farmers and traders. Amongst farmers, storage may have some added advantages because it increases market flexibility. Households with proper storage facilities do not need to market their produce immediately after harvest when prices tend to be low. They can store their produce and sell when prices are higher.

Most smallholder farmers do not have access to adequate storage infrastructure and end up selling their produce soon after harvest, also because they need the money involved. Smallholder farmers often rely on open-air storage (Gabre-Madhin, 2001). Due to lack of storage facilities, most smallholder producers are keen to sell produce almost immediately after harvest in order to ease congestion, leading them to sell their produce at lower prices (Wilson *et al.*, 1995).

Market infrastructure

Smallholder farmers are usually served by poor market infrastructure. In some instances, market infrastructure is unavailable and farmers sell from the back of their trucks (Makhura, 2001). These conditions are not conducive for fresh produce, contributing to perishability and loss of produce. Additionally, produce sold under poor market conditions may not be attractive to consumers, putting farmers at risk of losing customers. Fresh produce tends to have a limited shelf life, therefore, they cannot be stored for long periods (Van Tilburg, 2005). This implies that such produce needs to be processed or to be sold while it is still fresh. When selling them, it is important to be cautious of market place conditions to keep them fresh. Market infrastructure such as sheds and stalls in spot markets is crucial in maintaining freshness of agricultural produce (Wilson *et al.*, 1995).

Road infrastructure

Agricultural commodities must move from the farms where they are grown to the retail outlets where they are bought. Road infrastructure and transport availability have an influence on smallholder market participation, especially if they are located far from the consumption centres (Gabre-Madhin, 2001). According to Bachmann and Earles (2000), one of the most important constraints facing agricultural markets throughout sub-Saharan Africa is transport infrastructure and the need to reduce transport.

The majority of villages in rural areas are served by an inadequate and poorly maintained road network (Montshwe, 2006). The poor conditions of roads, which are often impassable during the rainy season, have an adverse effect on the transportation of the produce. As transport generally marks the passage from one stage of the post-harvest system to the next, if the roads are poorly developed, it becomes difficult to move produce from one stage to another (Goletti and Wolff, 1998). If roads are in bad condition, travelling time is long, implying that it will be difficult to sell fresh produce within the required time limit (Dijkstra *et al.*, 2001).

Market transport

It is difficult to transport produce to the market in time (produce spoilage and losses) if there is no reliable private form of transport, since public vehicles tend to be limited in the rural areas (Bachmann and Earles, 2000). In addition, unavailability of reliable transport will increase transport costs, which in turn increases transaction costs amongst smallholder farmers (Zaibet and Dunn, 1998).

In southern Africa, most smallholder farmers usually pack their goods (especially vegetables) in sacks, which are then transported to the market places using public transport (Jayne *et al.*, 2002). This leads to bruises and damage and, thus, drastically reduces the quality of the agricultural produce being transported. Some farmers use their own vehicles to get to the market centres. Makhura (2001) pointed out that these farmers with these assets are able to move around in search of more rewarding markets. In addition, those farmers stand a better chance of getting market information from different markets.

Value adding

According to Robbins (2005), prices of primary agricultural produce have fallen steeply, but retail prices for the same packaged, cut and processed products in industrial countries, have increased. This means that value adding activities can earn farmers additional income. Value adding can be in the form of grading, sorting, cutting, packaging in standard weights and processing of produce (Mather, 2005). Lack of value adding and agro-processing is part of missing markets amongst smallholder farmers in marketing. Agricultural produce from smallholder farmers usually are poorly packaged. With few exceptions, most smallholder farmers cannot add value to their produce because they do not know its importance and lack processing technology (Louw *et al.*, 2007). Inability to add value to agricultural produce by smallholder farmers excludes them from interesting markets.

Using the literature, certain hypotheses were drawn and variables were chosen to be included in the empirical model for the emerging and smallholder farmers in the Kat River Valley. These are summarised as:

- the availability of an extensive social capital structure has a positive influence on the market participation choice;
- the ability to add value exerts a positive impact on market participation choice;
- access to market information influences market participation choice positively;
- there is a positive relationship between extension services and marketing choices;
- households with expertise on grades and standards are expected to use formal markets than those without;
- farmers forming part of agricultural organisations and those involved in group marketing are more likely to market their produce using both formal and informal markets;
- marketing in both formal and informal markets increases with the availability of good road infrastructure;
- availability of reliable market transport positively influences marketing choices;
- existence of contractual agreements is hypothesised to impact positively on the formal channel choice;
- farmers are more likely to use formal and informal markets, with the availability of good market infrastructure.

3.3 The study area: Kat River Valley

Kat River Valley (Figure 3.2) is located in the Eastern Cape Province, the second largest of the nine provinces in South Africa (Ngqangweni, 1999). It is situated northeast of Grahamstown, in the foothills of the Winterberg and the Amatole Mountains (Magni, 1999). The Kat River Valley forms part of the Nkonkobe Local Municipality, which falls under the Amatole District Municipality. Before the change of government in 1994, the upper part of the Kat River Valley was part of the Ciskei homeland – one of several black racial reserves created during the apartheid era. The Kat River Valley is approximately 80 km in length and 1,700 km^2 in area (Motteux, 2001). Its catchment includes the areas of Seymour, Balfour, Fort Beaufort and other smaller rural communities. According to McMaster (2002), the Kat River Valley can be divided into three different sub regions (Figure 3.2), namely: the Upper Kat, Middle Kat and the Lower Kat.

According to Magni (1999), Kat River Valley's climate can be described as mild. The rainfall is unevenly distributed within the area. It ranges between 400 mm and 1,200 mm, where the least rainfall is received at the confluence with the Great Fish River and the highest, in

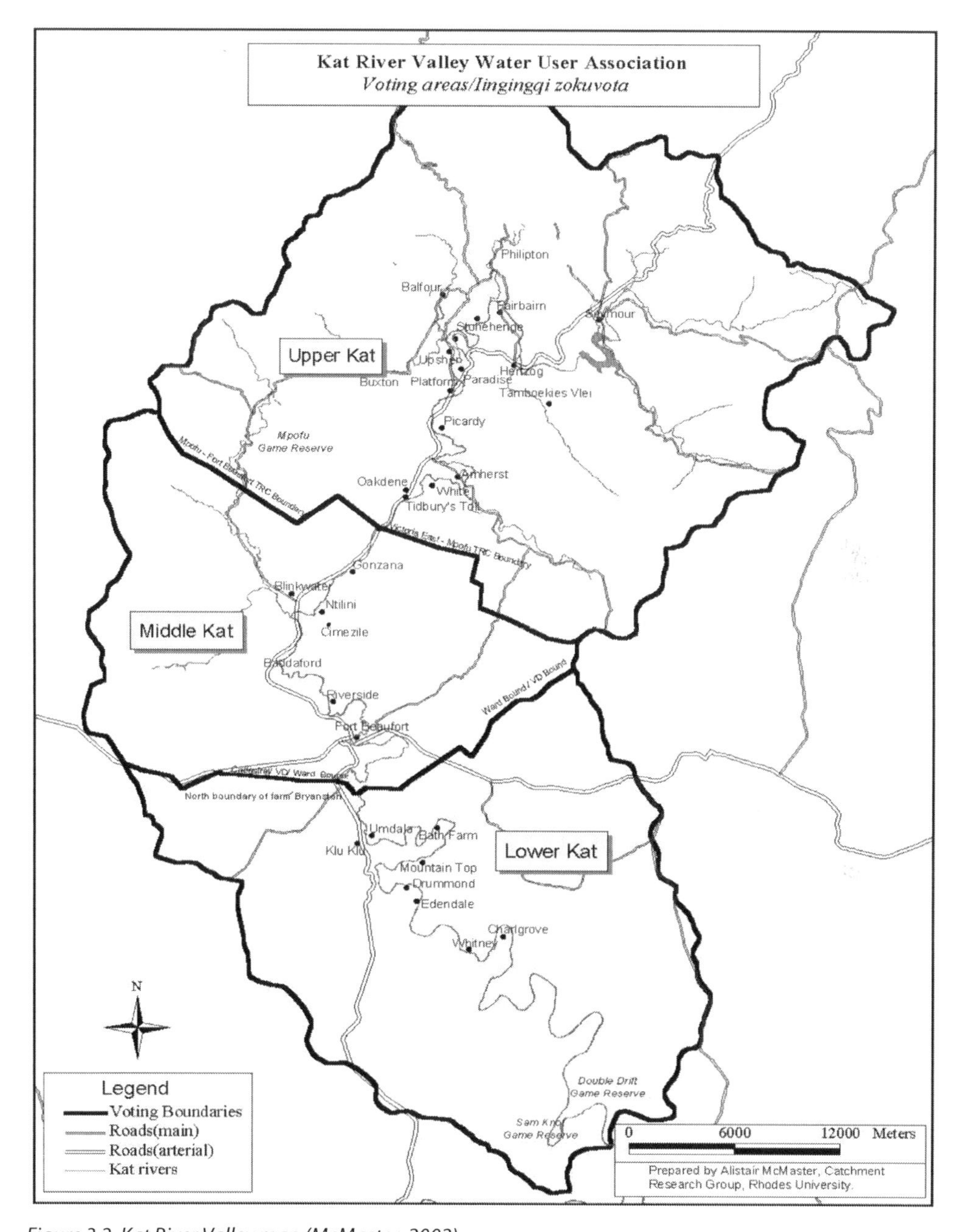

Figure 3.2. Kat River Valley map (McMaster, 2002).

the mountainous northern region of the catchment (Magni, 1999). Although the rainfall is relatively high in the mountainous region, much of the area in the catchment can be regarded as sub-humid to semi-arid. Kat River Valley receives both summer and winter rainfall. Approximately 75% of the mean annual precipitation is received between October/November and February/March, where the highest rainfall figures are recorded in March. The temperatures range from moderately hot summers to cool moderate winters (Motteux, 2001).

Kat River Valley is characterised by a variety of land uses, ranging from export-oriented citrus farming, commercially oriented rangeland stock farming to small-scale vegetable and crop production and stock farming (McMaster, 2002). Commercial farmers are mainly located in the Middle Kat and Lower Kat, whereas smallholders and emerging farmers mostly practice agriculture in the Upper catchment. According to Nel (1998), during apartheid, citrus farming was only practiced by white commercial farmers. As part of black empowerment, some black farmers were helped operate citrus farms by Ulimocor, after apartheid. In addition to citrus farming, Farolfi and Rowntree (2005) noted that vegetable gardening is an important occupation amongst emerging smallholder farmers. Most of these vegetables are grown on fertile plots lying adjacent to rivers and streams. Whereas some farmers practice sprinkler irrigation, irrigation by hand is practiced by farmers who lack irrigation infrastructure.

Population in the Kat River Valley is composed of different races, and the racial composition is the result of history and apartheid. The Upper and Middle Kat belong mainly to black *Xhosa* speakers and coloured Afrikaans speakers, whereas the Lower Kat belong mainly to white English speakers. Of the total population in the area, approximately 94.28% are black *Xhosa* people, 4.12% coloured and 0.76% white (Motteux, 2001).

3.4 Data collection and analysis

Data was collected from a sample of emerging and smallholder farmers who are producing a marketable surplus in the Kat River Valley. Farmers were stratified according to farming types: cattle, vegetable and citrus farmers. One hundred farmers were randomly selected for the survey, where 43 were cattle farmers, 43 being vegetable farmers and 14 (total population) citrus farmers. Data was collected from household heads through face-to-face interviews. In the absence of the head, the spouse or any family member who is directly involved in the farming activities and management was interviewed. The respondents provided demographic data, infrastructure development and transport availability data, amount of crop and livestock sold at the market, market proximity, market institutional arrangements and difficulties involved in market exchange. To analyse these data, descriptive and multivariate statistical analyses (the multinomial logistic regression model) were used to test the hypotheses. The main descriptive indicators that were employed were frequencies and mean values. The Statistical Package for Social Scientists (SPSS version 15.0) was used to generate frequency tables and to test the hypotheses.

The multinomial logistic regression model was used to test the institutional and technical factors that influence households from using greater depth marketing methods, which have the potential of increasing their incomes. Multinomial logistic regression can be used to predict a dependent variable, based on continuous and/or categorical independent variables, where the dependent variable takes more than two forms (Hill *et al.*, 2001). Logistic regression does not assume a linear relationship between the dependent variable and independent variables, but requires that the independent variables be linearly related to the logit of the dependent variable (Gujarati, 1992). Pundo and Fraser (2006) explained that the model allows for the interpretation of the logit weights for the variables in the same way as in linear regression.

The model has been chosen because it allows analysis of data where participants are faced with more than two choices. Smallholder and emerging farmers under study are faced with three choices regarding sales outlet, which are; formal market participation, informal market participation and non-market participation. Firstly, the farming households are assumed to decide whether to market their produce or not. When they choose to market, they then decide on the marketing channel to be used (either formal markets or informal markets). These decisions are made based on the option, which maximises their utility, subject to institutional and technical constraints.

As such, the utility maximising function can be given as:

$$\text{Max U} = U\,(C_k, R_{fk}, R_{ik}; H_u) \qquad (1)$$

Where:
Max U denotes the maximum utility that can be attained from agricultural production;
C_k represents the consumption of produced goods by the household;
R_{fk} represents revenue gained from formal market participation;
R_{ik} represents revenue gained from informal market participation;
H_u represents a set of institutional and technical factors shifting the utility function.

From the utility maximising function, it can be seen that households make decisions to produce, consume and market, subject to institutional and technical factors. Therefore, if the costs that are associated with using a particular channel are greater than the benefits, households will be discouraged from using it, shifting to another option that maximises their utility. For instance, if there are technical and institutional challenges specific to formal markets, that increase marketing cost above the revenue, households will be discouraged from using formal markets. They then analyse the costs associated with informal markets. If the institutional and technical factors that are unique to informal markets increase marketing costs above returns, then households will decide not to sell their produce. In this case, consumption of own production is considered as the option that provides the maximum utility.

In the utility function, the amount of good k that is consumed or sold does not have to exceed the amount that is produced. However, Hill *et al.* (2001) pointed out that it is difficult to measure utility directly; therefore, it is assumed that households make participation choices depending on the option that maximises their utility. Thus, decisions to participate in either formal or informal markets or even not participating signify the direction, which maximises utility. With the given assumption, multinomial regression was used to relate the decisions to participate in formal markets, informal markets or not participating and the factors that influence these choices.

A typical logistic regression model, which was used is of the form:

$$\text{Logit}(P_i) = ln\,(P_i / 1 - P_i) = \alpha + \beta_1 X_1 + \ldots + \beta_n X_n \qquad (2)$$

Where:
$ln\,(P_i / 1 - P_i)$ = logit for market participation choices;
P_i = not participating in markets;
$1 - P_i$ = participating in markets;
β = coefficient;
X represents covariates.

In the model, market participation choice, with three possibilities, *viz.* formal markets, informal markets and not participating in markets, has been set as the dependent variable. The variable of non-market participation was accepted as the baseline group; therefore, it took the value of zero. Informal market participation took the value of one and formal market participation was equated to two. The multinomial model was used to determine the odds of both formal market participation versus non-participation and informal market participation versus non-participation. It follows that P_i represents the probability of not participating in produce marketing and $(1 - P_i)$ represents the probability of either informal market participation or formal market participation.

In order to capture the dependent variable closely, households were asked the marketing channels they used for each transaction. The main advantage for using a transaction in analysis is its relationship to one type of channel (Dijkstra *et al.*, 2001). Thus, it is possible for a farmer to use more than one channel in marketing produce but it is impossible to carry out a transaction in more than one channel.

By fitting explanatory variables into the model, the model is presented as:

$$\begin{aligned} ln\,(P_i / 1 - P_i) = {} & B_0 + B_1 MKTINFO + B_2 GRDS + B_3 EXT + B_4 ORGMEM + B_5 MNGTYP \\ & + B_6 RDINFR + B_7 TRANS + B_8 ADDVAL + B_9 MKTINFR + B_{10} STOR \\ & + B_{11} CONTRCT + B_{12} SOVIALK + B_{13} PART + B_{14} TRAD + Et \end{aligned}$$

The variable reflecting access to market information (MKTINFO) was measured by the household's ability to get market information in time and the ability to interpret it correctly. In order to capture this variable closely, households were interviewed on the communication networks that are accessible to them. Ability to communicate in either English or Afrikaans was used to measure the accuracy of information interpretation by households. Access to information has been set as a dummy variable, where a household with access to information takes the value of one and a household that has no access to information takes a value of zero. Access to information was expected to influence market participation positively; implying that households with access to information are more likely to participate in marketing, making use of formal markets. Another variable that is closely linked to information availability is access to extension services (EXT) such as access to farming advice and knowledge through extension officers. This variable was also allocated dummy values where households with access to extension services took the value of one and zero if otherwise.

According to Reardon and Barrett (2000), smallholder farmers have difficulties in meeting market grades and standards, leading to exclusion of such farmers from participating in mainstream agriculture. Expertise on grades and standards (GRDS) was recorded in this instance to investigate whether it influences marketing participation choices or not. Households were asked if they were aware of market grades and standards, and whether there were problems meeting such standards. The households with knowledge on grades and standards, and had no problems meeting them were set to have expertise on grades and standards. Such households took the value of one and those households with no expertise on grades and standards were equated to zero. Households with expertise on grades and standards are more likely to use of formal markets than those without, thus an expected positive relation.

Collective action is important in agricultural marketing because it contributes towards reduced transaction costs and it strengthens farmers' bargaining and lobbying power (Kherallah and Kirsten, 2001). Collective action is measured by two main variables, which are organisational support services (ORGMEM), and group or individual participation (PART). Respondents were asked whether they belonged to an organisation or not and whether they sold output in groups or individually, the responses were allocated dummy values. Both the variables are anticipated to impact positively on market participation choice among the smallholder farmers.

The availability and condition of both road and market infrastructures are thought to have an influence on marketing efficiency. Where the infrastructure is unavailable or poor, farmers are discouraged from using it, thereby limiting market participation. Therefore, the availability of good road and market infrastructures are expected to exert a positive influence on market participation. Road infrastructure (RDINFR) is measured by the adequacy of the road networks that are accessible to households and their condition. Market infrastructure (MKTINFR) is measured by the availability of infrastructure, such as marketing stalls and their condition. Dummy values are used to define the variables, where in both cases, one indicates good condition and zero indicates either unavailability or poor condition.

Ownership of market transport (TRANS), specifically vehicles, was used to measure the availability of produce transportation facilities by households. Moreover, the availability of transportation facilities helps reduce long market distance constraint, offering greater depth in marketing choices. In cases where households owned a vehicle, the variable took the value of one, and zero if the household did not own any form of vehicle. This variable is expected to have a positive influence on the market participation choices.

Social capital (SOCIALK) refers to personal social networks that encourage market participation. It is through these networks that trust is developed, which, in turn, encourages cooperation and regular exchanges. Therefore, social networks reduce transaction costs, leading to diversified market participation choices. Again, information and production resources can be transmitted through these networks. In order to capture this variable, respondents were asked about the relationship with their customers and if any were regular customers. The availability of an extensive social capital structure is expected to impact positively on the dependent variable.

The availability of contractual agreements (CONTRCT) ensures the availability of a guaranteed market for the farmers, thus promoting market participation and including smallholder farmers in mainstream agriculture. In other words, the existence of a guaranteed market reduces the costs that are associated with searching for potential buyers, thereby encouraging participation in formal markets. This variable is expected to have a positive relationship with the dependent variable

The types of farming (FMNGTYP) have been divided into two dummy variables (FMNGTYP1 and FMNGTYP2). Under FMNGTYP1, one represents cattle farmers, otherwise zero, whereas, under FMNGTYP2, one represents vegetable farmers, and zero for farmers not producing vegetables. Another analysis, which excluded the citrus farmers (they are too few to be analysed in a separate group) was carried out to check the outcome for vegetable farmers and cattle farmers only. The type of farming variable was included into the model in order to capture the differences in the nature of produce from different farming types. Thus, in some types of farming, formal market penetration may be easier than in the other types. For instance, Matungul *et al.* (2002) pointed out that formal livestock markets are readily accessible to both commercial and small-scale livestock farmers in South Africa, owing to public investment in sales yards. The variable can take either a positive or a negative value.

The ability to add value to agricultural produce is captured by the variable (ADDVAL). Dummy values are used to define the variable where those households who add value to their produce, take the value of one and those who do not, equal to zero. It is hypothesised that the ability to add value exerts a positive impact on market participation. This positive relationship is because households with the ability to add value can sell their produce in an improved state, which can be more appealing to customers. The variable storage facilities (STOR), is closely related to value adding. Good storage facilities reduce loss of produce and urgency of produce selling, and maintain the physical state of produce. Thus, households with good storage facilities are more likely to participate in formal markets, hence a positive relationship.

Traditions and beliefs are part of informal institutions that can influence marketing choices. In the model, the dummy variable, guided by tradition and beliefs (TRAD) is used to determine effects of tradition and beliefs on marketing. Households were asked on the extent to which tradition plays a role in their agricultural activities. They were also asked if they were open to new farming and marketing methods offered by non-family members. The variable was allocated dummy values were households with strong traditional guidance took the value of one and zero if otherwise. The variable is expected to take a negative value because household depending on traditions and beliefs are normally not liable to change (Dorward *et al.*, 2003). Such households would rather stick to what they know especially if the marketing environment is changing rapidly. Thus, such households are less likely to participate in the formal markets.

3.5 Empirical results

The empirical results presented in this section give a summary of the descriptive results and the logistic regression results.

3.5.1 Descriptive results

The descriptive results for the demographic characteristics show that from all the interviewed respondents, there were a larger proportion of male respondents (58%) as opposed to females. However, there were greater proportions of females (63%) in vegetable farming, but in both cattle and citrus farming, there were greater proportions of males. A large number of females in vegetable farming can be explained by day-to-day vegetable supervision by females where men move to cities/town in search of jobs. The majority (72%) of the smallholder and emerging farmers in the Kat River Valley are above 49 years of age. The educational level among the sampled farmers is generally low, where 18% of the household heads never attended school and 39% have gone up to primary level. Household size ranges from a minimum of two people to a maximum of 18, with a mean of 7.1 in each household. According to Randela (2005), a larger household size has a negative effect in produce marketing because the household needs to supply household consumption before it decides to sell. Evidence from the research confirms this line of argument because larger households in this research sold less produce as compared to smaller households.

The land that is available to smallholder farmers in the Kat River Valley is usually shared between residential and farming purposes. This situation leaves less arable land for farming purposes. In addition, most smallholder farmers do not own the land they farm on, even though they have rights to use it. The sampled farmers had access to small pieces of land where 77% of the households had less than 2 hectares. Comparing land accessibility to different farming types, cattle farmers had the least land sizes where all of them had less than 2 hectares. Smaller areas of arable that is used by cattle farmers can be explained by use of communal grazing land for animal feeding. In such cases, it is difficult to measure the amount of communal grazing land that is available to each household.

Household incomes of the respondents are received from five main sources, which are farming, wages, pensions, social grants, and other small household business activities. Of importance, is the fact that in the absence of pension and social grants, 74% of the households view farming as their main source of income. When selling produce, households had different reasons for choosing the marketing channels they use. A summary of the main market outlets used by smallholder and emerging farmer are presented in Table 3.1.

The markets outlets shown on Table 3.1 were divided into formal and informal markets, where informal markets embrace unofficial transactions between farmers and traders, and farmers and consumers whereas formal markets are guided by formal rules and regulations. As shown in Table 3.1, more vegetable and cattle farmers make use of informal markets than formal markets. The marketing channels used by citrus farmers present a slightly different trend compared to that of vegetable and cattle farmers. Even though all citrus farmers mainly use export markets, between 20 and 40% of their produce is sold in local informal markets. In an effort to find the reasons why the sampled farmers preferred the markets they use, they

Table 3.1. Main market outlets used by sampled households.

Type of farming	Main market outlets	Market type
Vegetables (n=43)	farm gate (46.5%)	informal
	Fort Beaufort (32.5%)	formal and informal
	around the village (14.0%)	informal
	roadside (7.0)	informal
Citrus (n=14)	export markets (100%)	formal
Cattle (n=43)	private sales (39%)	informal
	speculators (26%)	informal
	auctions (23%)	formal
	abattoirs (9%)	formal
	butcheries (3%)	formal

were interviewed on the prevailing marketing problems they faced during selling and those that constrained them from manoeuvring into more rewarding marketing arrangements. The most frequently mentioned marketing challenges are shown in Table 3.2.

Farmers were asked to clarify the challenges listed in Table 3.2. For example, farmers were asked what they meant by market information. The most mentioned answers were related to information on the prevailing prices, type of goods required in the markets and alternative markets. Farmers explained that the information to which they had access was unreliable because they usually got it from either other people in the village who are involved in selling or from the rural traders. They pointed out that they rarely trusted such information but they had no option because those are the only sources accessible to them. Alternatively, the farmers explained that they would just take chances and go to the market place without any prior information to receive the same price as other people selling at that selling point.

Taking a closer look at the marketing challenges that were cited by households, they could all be resolved through technological and institutional innovation. For instance, low prices for produce which can be related to poor produce quality, inability to reach other markets, being uninformed and abundance of the same produce in the markets. Poor produce quality can be resolved by availability of information on grades and standards and an improvement in technology for storage and transportation. Farmers may be able to reach other markets if they participate in groups because they share information and broaden social capital within the groups. In addition, when farmers market in groups, they eliminate competition and may diversify into producing other crops, reducing market pressure.

Table 3.2. Marketing challenges among sampled households.

Rank	Marketing challenges	Households affected (%)
1	Lack of capital	80
2	Bad roads	79
3	Low prices for produce	67
4	No reliable markets	65
5	No market information	64
6	Lack transport/high transportation costs	59
7	No exposure to other markets	53
8	Lack of storage facilities	51
9	High competition	46
10	No convenient place to sell from	38
11	No transparency in marketing channels	18

3.5.2 Variable correlations

A correlation analysis was run to determine the significant relationships between the explanatory variables that were used in the model. The variables correlations were tested using the Pearson correlation, at the 5% significance level.

Based on the results, it was noticed that some of the variables were correlated. Extension contact was significantly related to market information, organisational support services, expertise on grades, ability to add value and availability of an extensive social capital. The other variables that were correlated included contractual agreements and group participation, storage facilities and expertise on grades, availability of an extensive social capital and group participation, and market information and expertise on grades. Since the variables were highly correlated (where $r>0.5$ in most cases), it raised suspicion of the multicollinearity problem. The problem may have been caused by the use of dummy variables as Rao and Rao (1998) stated that multicollinearity is common when using dummy variables. However, Hill *et al.* (2001) explained that multicollinearity does not influence the reliability of the model as a whole but affects the individual predictor variables.

In order to rectify the problem and improve the results, one of the explanatory variables was dropped. The extension service variable was chosen because most variables were correlated to it. It seemed reasonable to drop the variable because not much information would be lost after dropping it. The other variables would be used to capture its importance because most of the knowledge related to market information, grades, adding value and importance of group support was imparted through extension workers. After dropping the extension service variable there was minimal correlation of variables, with only two variables being significantly correlated. A reduction in the correlation of explanatory variables indicates that the multicollinearity problem was corrected. The remaining explanatory variables were then used in the model.

3.5.3 Significant variables in the model

The hypotheses generated in Section 3.5 were tested in a multinomial logistic regression model. The results of the parameter estimation in the model are presented in Table 3.3.

The table shows the estimated coefficients (β values), standard error, significance values and odds ratios of variables in the model. According to Gujarati (1992), the coefficient values measure the expected change in the logit for a unit change in the corresponding independent variable, other independent variables being equal. The sign of the coefficient shows the direction of influence of the variable on the logit. It follows that a positive value indicates an increase in the likelihood that a household will change to the alternative option from the baseline group. On the other hand, a negative value shows that it is less likely that a household will consider the other alternative (Gujarati, 1992; Pundo and Fraser,

Table 3.3. Multinomial logistic results for informal and formal market choices as compared to non-marketing choice (n=100).

Variable	Informal market choice				Formal market choice			
	Coefficient	Std. Error	Significance[b]	Odds ratio	Coefficient	Std. Error	Significance[b]	Odds ratio
MKTINFO	2.686	1.050	0.011*	14.673	4.217	1.385	0.006*	67.83
GRDS	-1.623	0.905	0.073	0.197	3.830	0.848	0.016*	46.06
ORGMEM	0.788	0.786	0.316	2.199	1.324	1.384	0.330	3.758
FMNGTYP1	-0.248	0.754	0.742	0.780	-1.164	1.854	0.530	0.312
FMNGTYP2	1.798	0.854	0.028*	6.038	-1.543	1.017	0.033*	0.214
FMNGTYP1[a]	-0.523	0.675	0.065	0.593	0.762	0.545	0.048*	2.143
RDINFR	0.862	0.841	0.305	2.368	2.992	2.171	0.168	19.92
SOCIAILK	0.222	0.948	0.050*	1.248	1.180	2.863	0.031*	3.254
ADDVAL	1.352	1.079	0.210	3.865	0.392	0.218	0.860	1.479
MKTINFR	2.557	1.030	0.013*	12.897	-0.687	0.026	0.735	0.503
STOR	0.584	0.777	0.453	1.793	0.259	1.873	0.890	1.296
CONTRCT	0.844	0.755	0.263	2.326	2.803	1.912	0.047*	16.49
TRANS	0.843	0.774	0.276	2.323	0.449	1.644	0.785	1.567
PART	1.899	0.854	0.026*	6.679	1.997	1.418	0.039*	7.367
TRAD	2.477	1.441	0.031*	11.905	-2.144	2.296	0.007*	0.117
INTERCEPT	-4.934	2.860	0.048	-	-17.069	4.466	0.044	-
Goodness-of-fit								
	Chi^2	df	Significance					
Pearson	111.372	168	0.150					
Deviance	84.301	168	0.988					

[a] Variable tested for 86 respondents (43 vegetable farmers and 43 cattle farmers).
[b] * Statistically significant at 5% significance level.

2006). Therefore, a positive value implies an increase in the likelihood of changing from not participating in marketing to either informal or formal market participation choice.

The variables were tested at the 5% significance level. Thus, if the significance value is greater than 0.05, then it shows that there is insufficient evidence to support that the independent variable influences a change away from the baseline group and *vice versa*. The odds ratio indicates the extent of effect on the dependent variable caused by the predictor variables. Its value is obtained by calculating the anti-logarithm of each slope coefficient of predictor variables. A value greater than one implies greater probability of variable influence on the logit and a value less than one indicates that the variable is less likely to influence the logit. The standard error measures the standard deviation of the error in the value of a given variable (Gujarati, 1992; Hill *et al.*, 2001).

As indicated in Table 3.3, some predictor variables influence market participation choices significantly. Of the 14 independent variables used in the model, five and six variables in informal and formal market choices respectively, are statistically significant at the 5% significance level. In all but one of the cases, the signs of the estimated coefficients are consistent with the *a priori* expectations.

Access to market information has a positive sign for both formal and informal market choices, which is consistent with the *a priori* expectations. The significance values of 0.011 for the informal market choice and 0.006 for the formal market choice imply that there is enough evidence to support that an increase in the availability of market information results in an increase in both informal and formal market participation. The larger values in odds ratios show that households are most likely to increase participation in both informal and formal markets with the availability of market information. As shown by the coefficients, the increase in formal marketing, resulting from market information availability is about twice the increase in informal marketing.

Expertise on grades and standards is significant for the formal market choice with a significance value of 0.016. A positive sign on its coefficient indicates that an improvement in expertise on grades and standards results in an increase in the formal market participation choice by households. When households acquire expertise in grades and standards, they prefer selling their produce in the better paying formal markets, in order to cover costs associated with acquiring the expertise (Reardon and Barrett, 2000).

A positive and significant (0.047) relationship was found between formal market participation and the availability of contractual agreements. The relationship implies that households tend to increase formal market participation with the availability of contractual agreements. This relationship is most likely due to the influence of the citrus farmers. The value of the odds ratio (16.49) supports the higher probability of the variable influence on the formal market choice.

The variable existence of extensive social capital is significant for both informal (0.050) and formal (0.031) market choices. The positive relationship in both formal and informal market participation choices explains that an increase in social capital results in households shifting from non-participation to formal and informal market participation. The odds ratios for both formal and informal market choices suggest a higher probability of shifting to formal and informal marketing with an increase in social capital. Therefore, it can be concluded that social networks are important in produce marketing, regardless of the choice of market being used.

It was expected that the availability of good market infrastructure could have a positive influence on alternative market participation choices, away from not participating in marketing. However, the *a priori* expectations hold true for the informal market choice only. There is sufficient evidence (significance value of 0.013) to support that the availability of good market infrastructure is likely to encourage households to market their produce through informal channels. Unlike formal channels where market infrastructure is not important for farmers, as they supply their produce in bulk once harvested to the higher level of the marketing channel (Takavarasha and Jayne, 2004).

Group participation in marketing was expected to have a positive influence on the dependent variable. The results shown in Table 3.3 for this variable are consistent with the *a priori* expectations. For both formal and informal market choices, there is enough evidence to support that when households market their produce in groups, there is a higher chance of participating in either formal or informal markets. Thus, group participation encourages market penetration among smallholder farmers who find it difficult individually to gain market access.

A positive and significant (0.031) relationship was found between informal marketing and guidance from traditions and beliefs. These results are contradiction to the *a priori* expectations, where guidance from traditions and beliefs was expected to influence the dependent variable negatively. The positive relationship between the variables may possibly be explained by traditional wisdom and skills passed on in families and creation of marketing links through traditions and beliefs. For instance, some households may prefer to sell their produce (especially in cattle marketing) to people they are familiar with. On the other hand, there is a negative and significant (0.007) relationship between formal marketing and guidance from traditions and beliefs. The explanation to this relationship may be that the marketing environment is ever changing (Kherallah and Kirsten, 2001); therefore, if farmers are to be part of the formal markets, they have to be receptive to changes.

The variable FMNGTYP2 is positively related to informal marketing but negatively related to formal marketing. The relationships imply that there is a difference between farming types and marketing in both formal and informal markets. The results show that vegetable farmers, when compared to a combination of citrus and cattle farmers, are more likely to use informal marketing channels than formal marketing channels. When the farming type variable was tested for 86 respondents (43 cattle and 43 vegetable farmers), a positive relationship was noted for the formal marketing choice. These results explain that cattle farmers are most likely to use the formal marketing channels as compared to vegetable growers in the Kat River Valley.

The goodness-of-fit test for a logistic regression model measures the suitability of the model to a given data set. An adequate fit corresponds to a finding of non-significance for the tests (Hill *et al.*, 2001). The results for the goodness-of-fit test shown in Table 3.3 indicate that the model fits the data well. Thus, the results for both Pearson and Deviance chi-squared methods show that the multinomial logistic regression model is well suited to predict the influence of independent variables on the dependent variable.

3.6 Conclusions and recommendations

This research has attempted to identify factors influencing market participation choices among smallholder and emerging farmers in the Kat River Valley. Evidence from the research supports literature that pointed out that smallholder and emerging farmers usually use informal markets in selling their produce. Furthermore, there are some challenges in the market environment that discourage these farmers from using formal markets. The statistically significant variables, at 5% level are access to market information, expertise on grades and standards, availability of contractual agreements, existence of extensive social capital, availability of good market infrastructure, group participation and reliance on tradition.

The results suggest several ways in which smallholder farmers can actively market their produce. The findings suggest that an adjustment in each one of the significant variables can significantly influence the probability of market participation. That is, technological growth and institutional developments that affect such variables can help farmers improve participation and encourage formal market participation. Possible adjustment paths are explained in the Section 3.6.1 'Policy recommendations'.

3.6.1 Policy recommendations

With regard to the smallholder and emerging farmers' marketing challenges revealed by the empirical results, policy recommendations can be suggested. This section gives a series of options that can be considered in South Africa, in an effort to help smallholder and emerging farmers reach their full potential.

Encourage collective action through formation and consolidation of farmer groups

Literature has revealed that agricultural produce is being distributed through organised marketing channels, away from spot markets. On the other hand, the study has shown that emerging and smallholder farmers have problems in accessing the formal markets individually, partly because of relatively small marketable surpluses, high transaction costs and problems in meeting grades and standards. Given such information, it is important to establish the suitability of collective action (in the form of farmer organisations or trade associations) as an institutional vehicle for linking smallholder farmers to agribusiness supply chains. The public sector can promote collective action by creating a legal framework, which is favourable for registration and operation of farmer organisations. Collective action can be considered because it strengthens smallholders' market position, bargaining power and lobbying power (Kherallah and Minot, 2001). In addition, fixed transaction costs can be borne by more shoulders, resulting in a decrease in individual costs. In addition, through shared knowledge, farmers can ensure market grades for produce, within the producer organisations. However, if collective action is not managed properly, it often leads to problems of free-riding and opportunistic behaviour, where some individuals take advantage of the coordination of others, without personally investing their efforts. These actions may be a disincentive to some members, which may result in failure to reach goals, and even in coordination break up (Olson, 2009). Farmer groups can only be successful is they are based on trust, honesty, mutual respect and commitment to the group (Masuku *et al.*, 2003). This brings out the suggestion that when choosing group members, notably farmers who are working towards the same goal should cooperate. In addition, rules and roles within the group ought to be specified from the beginning, and effective governance and management practices adopted. Such practices, together with technical and financial support, and farmer training will help farmer groups develop into business-oriented entities.

Encourage farmer training

Some small-scale famers prefer sticking to traditional ways of producing and marketing, and often are excluded from dynamic formal markets. If these farmers have to become part of the formal agribusiness chain, intervention should start at an individual farmer level. Farmers need to strengthen traditional knowledge with training, particularly, training in management and technical skills that are required by smallholders in order to master the commercial requirements of producing for a competitive market, and be able to manage business risks and product innovations. Even where small-scale farmers have formed farmer or trading organisations, training is important for modernising their value chain knowledge and for improving their management and business alliance practices.

Promote contract farming

Contract farming is important to both the farmers and the contractors because it ensures a market for produce and supplies to the contractors. However, to get contractual deals, farmers should be able to provide a relatively larger output. When smallholder farmers operate in producer groups, they may be able to increase their output and be part of the contractual deals. The public and private sectors can help to facilitate contractual arrangements, but the farmers have to be willing to cooperate. Once they get contractual agreements, an entrepreneurial culture can be developed, where farmers produce for marketing, rather than trying to market what they have produced. Again, it is critical to develop trust between the farmers and the contractors, even though it should be supported by legal compliance.

Farmers can gain trust by delivering the required produce in time and contractors can develop trust by paying the producers in time according to the contract conditions. Such an environment encourages marketing and is advantageous to both parties.

Ensure the availability of market information to emerging and smallholder farmers

It has been highlighted that access to timely market information is still a problem among the smallholder farmers. As such, market information should be consistently supplied to the farmers through both private and governmental organisations. In an effort to make information available, it is important to know the types of market information that is necessary for different markets or market segments, such as specific rules, pricing, grades and standards; and train the farmers on how to use the information. Of equal importance, is devising the ways of disseminating the information, in order to reach all interested smallholder farmers. When devising these ways, it is important to consider the non-homogeneity of smallholder and emerging farmers, in terms of education, location and the availability of communication assets. For example, radio programs conducted in different languages and farmer workshops can be considered for information dissemination.

Encourage value adding activities

The farmers indicated that they do not know the importance of value addition, which is the reason why they are not involved in such practices. Therefore, knowledge related to value adding should be disseminated to the farmers, because value adding can open up opportunities and increase the farmers' profitability. It is important that the farmers, in cooperation with the private and public sectors, develop and initiate value-adding practices. Practices that do not need a lot of capital like packaging, cutting and drying can be considered by the farmers without outside help. The private and public sector can assist with training the farmers about value adding and provide financial assistance especially for the practices that require larger capital commitments, such as processing.

Invest in rural infrastructure

The public and private sectors, and Public-Private Partnership (PPP) can support the emerging and smallholder farmers through technical innovations. These may be in the form of investments in infrastructure such as improved roads, telecommunications and market-related infrastructure. Development of such facilities can reduce transaction costs and induce farmers to move towards a commercial agriculture system (Van Tilburg, 2005). Local economic development programmes in South Africa can be used to support rural infrastructure development. In supporting infrastructure development through PPP, the roles of the state, and those of the private sector need to be defined clearly in all the stages of program design, implementation and funding. Use of funds also needs to be monitored at all stages. The smallholder and emerging farmers still have to play a role to ensure that the infrastructural facilities are provided for them. They may form an association to lobby for and represent them.

Stimulate government support policies for the rural areas

The smallholder and emerging farmers in South Africa are facing unfair competition in markets, from the formerly supported commercial farmers. The smallholders have to compete for a market share with the commercial farmers who are already equipped in marketing. In addition, they are facing competition from internationally imported produce. For example, due to subsidy policies in developed countries, cheap produce is imported into South Africa. In order to withstand both local and international competition, there is need for changes in policy frameworks and institutional arrangements in favour of the small-scale farmers. There is need for campaign for the elimination of trade distorting subsidies and the removal of barriers for agricultural exports from developing countries. The South African policy needs to be flexible and be able to support initiatives that work for small-scale farmers, in order to assist the small-scale sector. The state can also offer preferential access to markets for Farmer Organisations and small farmer trade associations.

Acknowledgements

This research formed part of the For Innovative and Regional Collaborative Project In support of the Small-scale Farmers Development (FIRCOP) collaborative research project funded by the Fund for Countries having Priority for Solidarity [FSP] of the Government of France, Grant Contract No. FIRCOP/GC/006/06. The financial support of this research is gratefully acknowledged.

References

Bachmann, J. and R. Earles, 2000. Postharvest handling of fruits and vegetables, ATTRA Horticulture Technical Note. 19 pp. Available at: http://attra.ncat.org/attra-pub/PDF/postharvest.pdf.

Benfica, R., D. Tschirley and L. Sambo, 2002. Agro-industry and smallholder agriculture: institutional arrangements and rural poverty reduction in Mozambique. Journal of Modern African Studies 39: 333-358.

Carrè, M. and D. Drouot, 2002. Rhythm versus nature of technological change. available at: http://www.crest.fr/doctravail/document/2002-56.pdf.

Carter, M.R. and J. May, 1999. Poverty, livelihood and class in rural South Africa. World Development 27: 1-20.

Delgado, C., 1999. Sources of growth in smallholder agriculture in sub-Saharan Africa: the role of vertical integration of smallholders with processors and marketers of high-value added items. Agrekon 38: 165-189.

Dijkstra, T., M. Meulenberg and A. Van Tilburg, 2001. Applying marketing channel theory to food marketing in developing countries: a vertical disintegration model for horticultural marketing channels in Kenya. Agribusiness 17: 227-241.

Dorosh, P. and S. Haggblade, 2003. Growth linkages, price effects and income distribution in sub-Saharan Africa. Journal of African Economies 12: 207-235.

Dorward, A and J. Kydd, 2006. Making agricultural market systems work for the poor: promoting effective, efficient and accessible coordination and exchange. Project Report. Preparation of the DFID RNRA Team Working Paper 2: Making Markets Work for the Poor. DFID, London, UK.

Dorward, A., N. Poole, J. Morrison, J. Kydd, and I., Urey, 2003. Markets, institutions and technology: missing links in livelihoods analysis. Development Policy Review 21: 319-332.

Eggertson, T., 1990. Economic behaviour and institutions: Cambridge Survey of economic literature. Cambridge University Press, Cambridge, UK.

Fafchamps, M., 1996. The enforcement of commercial contracts in Ghana. World Development 24: 427-448.

Farolfi, S. and K. Rowntree, 2005. Negotiation and game theory for water management in the Kat Basin, South Africa: development of a cooperative game theory model and a comparison with a role-playing game results derived of a negotiation approach. Available at: http://www.ceepa.co.za/docs/Desole.pdf.

Fenwick, L.J. and M.C. Lyne, 1999. The relative importance of liquidity and other constraints inhibiting the growth of small-scale farming in KwaZulu-Natal. Development Southern Africa 16: 141-155.

Food and Agriculture Organisation (FAO), 2004. Changing patterns of agricultural trade. The evolution of trade in primary and processed agricultural products. Available at: http://www.fao.org/docrep/007/y5419e/y5419e05.htm.

Frank, S.D. and D.R. Henderson, 1992. Transaction costs as determinants of vertical coordination in the US food industries. American Journal of Agricultural Economics 74: 941-950.

Gabre-Madhin, E.Z., 2001. Understanding how markets work: transaction costs and institutions in the Ethiopian grain market. IFPRI Research Report 124. International Food Policy Research Institute, Washington, DC, USA.

Goletti, F. and C. Wolff, 1998. The impact of post-harvest research. MSSD Discussion Paper 29. International Food Policy Research Institute, Washington, DC, USA.

Gujarati, D., 1992. Essentials of econometrics. McGraw-Hill, New York, NY, USA.

Hill, R.C., W.E. Griffiths and G.G. Judge, 2001. Econometrics (2nd edition). John Wiley and Sons, New York, NY, USA.

Hobbs, J.E., 1997. Measuring the importance of transaction costs in cattle marketing. American Journal of Agricultural Economics 79: 1083-1095.

Jaffee, S. and J. Morton, 1995. Marketing Africa's high-value foods: comparative experiences of an emergent private sector. Kendall/Hunt Publishing, Dubuque, IA, USA.

Jayne, T.S., J. Govereh, A. Mwanaumo, J. Nyoro and A. Chapoto, 2002. False promise or false premise: the experience of food and input marketing in Eastern and Southern Africa, perspectives on agricultural transformation: a view from Africa. Nova Science, New York, NY, USA.

Key, N. and D. Runsten, 1999. Contract farming, smallholders, and rural development in Latin America: the organization of agro-processing firms and the scale of outgrower production. World Development 27: 381-401.

Kherallah, M. and J. Kirsten, 2001. The new institutional economics: applications for agricultural policy research in developing countries. Markets and Structural Studies Division, Discussion paper 41, International Food Policy Research Institute, Washington, DC, USA.

Kherallah, M. and N. Minot, 2001. Impact of agricultural market reforms on smallholder farmers in Benin and Malawi. IFPRI Collaborative Research Project. International Food Policy Research Institute, Washington DC.

Lewis, W.A., 1954. Economic development with unlimited supplies of labour. The Manchester School 22: 139-191.

Louw, A., Madevu, H., Jordaan, D. and Vermeulen, H., 2007. South Africa. In: B. Vorley, A. Fearne and D. Ray, (eds) Regoverning markets: a place for small-scale producers in modern agrifood chains? Gower Publishing Ltd, Oxford, UK, pp. 73-81.

Lyster, D.M., 1990. Agricultural marketing in KwaZulu: a farm-household perspective. MSc Agric thesis, University of Natal, Pietermaritzburg, South Africa.

Machethe, C.L., 2004. Agriculture and poverty in South Africa: can agriculture reduce poverty? Available at: http://cfapp1-docs-public.undp.org/eo/evaldocs1/sfcle/eodoc 357114047.pdf.

Magingxa, L. and A. Kamara, 2003. Institutional perspectives of enhancing smallholder market access in South Africa. Paper presented at the 41st Annual Conference of the Agricultural Economic Association of South Africa, Pretoria, South Africa.

Magni, P., 1999. Physical description of the Kat River Valley. Department of Geography, Rhodes University, Grahamstown, South Africa.

Makhura, M.T., 2001. Overcoming transaction costs barriers to market participation of smallholder farmers in the Northern Province of South Africa. PhD thesis, University of Pretoria, Pretoria, South Africa.

Mangisoni, J., 2006. Markets, institutions and agricultural performance in Africa. ATPS Special Paper Series No. 27, African Technology Policy Studies Network, Nairobi, Kenya.

Mather, C., 2005. SMEs in South Africa's food processing complex. Working Paper No 3. Available at: http://www.tips.org.za/node/394.

Masuku, M.B., J.F. Kirsten, C.J. Van Rooyen and S. Perret, 2003. Contractual relationships between small-holder sugarcane growers and millers in the sugar industry supply chain in Swaziland. Agrekon 42: 183-199.

Matungul, P.M., G.F. Ortmann and M.C. Lyne, 2002. Marketing methods and income generation amongst small-scale farmers in two communal areas of KwaZulu-Natal, South Africa. Paper presented at the 13th International Farm Management Congress, Wageningen, the Netherlands.

McMaster, A., 2002. GIS in participatory catchment management: a case study in the Kat River Valley, Eastern Cape, South Africa. MSc thesis, Rhodes University, Grahamstown, South Africa.

Mellor, J.W. 1966. The economics of agricultural economics. Cornell University Press, Ithaca, NY, USA.

Meyer, N., T. Fenyes and D. Louw, 1998. The impact of market reform and land reform on regional trade and food security in the South African grain market. In: J. Van Rooyen, J.A. Groenewald, S. Ngqangweni and T. Fenyes (eds) Agricultural policy reform in South Africa. Africa Institute for Policy Analysis, Pretoria, South Africa, pp. 246-261.

Minot, N. and F. Goletti, 1997. Impact of rice export policy on domestic prices and food security. markets, trade, and institutions. International Food Policy Research Institute (IFPRI). Available at: http://66.102.9.104/search?q=cache:GNwDIZLsjh YJ:www.ifpri.org/divs/mtid/mtidr.asp+Goletti+-+institutions&hl=en&ct=clnk&cd=2&gl=za.

Montshwe, B.D., 2006. Factors affecting participation in mainstream cattle markets by small-scale cattle farmers in South Africa. MSc Agric thesis, University of Free State, Bloemfontein, South Africa.

Motteux, N., 2001. The development and co-ordination of catchment fora through the empowerment of rural communities, catchment research group. WRC Report K5/1014, Water Research Commission, Pretoria, South Africa.

Nel. E.L., 1998. An evaluation of community driven economic development, land tenure and sustainable environmental development in the Kat River Valley. Report to the Programme for Human Needs, Resources and the Environment. HSRC Publishers, Pretoria, South Africa.

Nel, E. and J. Davies, 1999. Farming against the odds: an examination of the challenges facing farming and rural development in the Eastern Cape province of South Africa. Department of Geography, Rhodes University, Grahamstown, South Africa.

Ngqangweni, S.S., 1999. Rural growth linkages in the Eastern Cape province of South Africa. MSSD Discussion Paper No. 33. International Food Policy Research Institute, Washington, DC, USA.

North, D., 1990. Institutions, institutional change and economic performance. Cambridge University Press, Cambridge, UK.

Olson, F., 2009. United Producers Inc. Chapter 11 Restructuring. Journal of Cooperatives 23: 130-140.

Poulton, C., A. Dorward, J. Kydd, N. Poole and L. Smith, 1998. A new institutional economics perspective on current policy debates. In: A. Dorward, J. Kydd, and C. Poulton, (eds.), Smallholder cash crop production under market liberalization: a new institutional economics perspective. CAB International, Wallingford, UK.

Pundo, M.O. and G.C.G. Fraser, 2006. Multinomial logit analysis of household cooking fuel choice in rural Kenya: the case of Kisumu district. Agrekon 45: 24-37.

Randela, R., 2005. Integration of emerging cotton farmers into the commercial agricultural economy. PhD thesis, University of the Free State, Bloemfontein, South Africa.

Rao, C.R. and M.B. Rao, 1998. Matrix algebra and its applications to statistics and econometrics (1st edition). World Scientific Publishing, Singapore, Malaysia.

Reardon, T. and C. Barrett, 2000. Agro-industrialization, globalization, and international development: an overview of issues, patterns, and determinants. Journal of Agricultural Economics 23: 195-205.

Robbins, P., 2005. The new trade environment and the plight of small holder farmers. Available at: http://www.ciat.cgiar.org/agroempresas/pdf/ctamisace2005proceedings/session2pages2346.pdf.

Ruijs, A., 2002. cereal trade in developing countries: a stochastic equilibrium analysis of market liberalisation and institutional changes in Burkina Faso. PhD thesis, University of Groningen, the Netherlands.

Sasseville, D.N., 1988. Harvesting and handling produce: plan now for high quality. Missouri Farm (May-June): 19-21.

Shiferaw, B., G. Obare and G. Muricho, 2006. Rural institutions and producer organizations in imperfect markets: experiences from producer marketing groups in semi-arid Eastern Kenya. IFPRI Working paper 60, International Food Policy Research Institute, Washington, DC, USA.

Takavarasha, T. and T.S. Jayne, 2004. Toward improved maize marketing and trade policies to promote household food security in Southern Africa. Available at: http://www.aec.msu.edu/maizemarket/JayneTakavarashaSummary.pdf.

Todaro, M.P., 1997. Economic development (6th edition). Longman, London, UK.

Van Tilburg, A., 2005. Food marketing system, market institutions and co-ordination roles. Wageningen University, the Netherlands. Market Information systems and agricultural commodities exchange. Available at: http://www.ciat.cgiar.org/agroempresas/pdf/ctamisace2005proceedings/session2pages2346.pdf.

Wynne A.T. and M.C. Lyne 2003. Rural economic growth linkages and small-scale poultry production: a survey of poultry producers in KwaZulu-Natal. Paper presented at the 41st Annual Conference of the Agricultural Economic Association of South Africa, Pretoria, South Africa.

Wilson, L.G., M.D. Boyette and E.A. Estes, 1995. Postharvest handling and cooling of fresh fruits, vegetables and flowers for small farms. North Carolina Cooperative Extension Service, Raleigh, NC, USA.

Zaibet, L.T. and E.G. Dunn, 1998. Land tenure, farm size, and rural market participation in developing countries: the case of the Tunisia olive sector. Economic Development and Cultural Change 46: 831-848.

4. Technical constraints to market access for crop and livestock farmers in Nkonkobe Municipality, Eastern Cape province

Ajuruchukwu Obi and Peter Pote

4.1 Introduction and problem context

With democratic rule in South Africa, policies were introduced to redress the extreme inequalities in income, wealth and livelihoods engendered by apartheid rule. There was the expectation that enhanced access to productive resources such as land and technical support would translate into increased agricultural productivity for the black farmers who make up the bulk of the smallholders in the country. According to Makhura and Mokoena (2003), the result of this would be increased incomes, which would contribute to poverty reduction. Recent studies suggest that this goal has not been realised and that there has rather been a growing pauperisation of the citizens, especially the black population, manifested in deteriorating unemployment rates and poverty levels (Klasen, 1997; Klasen and Woolard, 2005, May, 1998; UNDP, 2003, 2007). For instance, while the broadly defined unemployment rates in the country stood at about 31% in 1993 (on the eve of the inception of majority rule in 1994), they had deteriorated to about 38% by 1997, rising to about 39% in 2005. The indication is that the picture is unchanged in 2008 and 2009 going by the spate of protests for increased wages that have rocked the country of late. The provincial data are equally disturbing, according to studies conducted in the late 1990s and early 2000's which suggested that provincial unemployment rates in the Eastern Cape may have been in the order of 30-70% (Department of Labour, 2003; May, 1998).

As many studies show, poverty is closely related to unemployment, among other factors (Klasen, 1997). Expectedly, the Eastern Cape Province which has the highest unemployment rate in South Africa also has the highest poverty rates, estimated at 71% in 1998 (May, 1998). Data generated by the Department of Labour (2003) and the Development Bank of Southern Africa (2005), suggest that the situation could be worsening. According to available data, this high poverty rate in the country is accompanied by the highest levels of income inequality in the world (HSRC, 1996; Klasen, 1997; Lam, 1999; UNDP, 2007). According to the UNDP (2007), the Gini coefficient estimated for South Africa for 2006 stood at about 0.59. Such a result is consistent with the fact that, among the Medium Human Development countries to which South Africa is placed by the UNDP, it is one of the few whose Human Development Indices actually deteriorated since the early 1990s, having fallen from 0.735 in 1990 to 0.653 in 2004 (UNDP, 2006).

Analyses based on comparable consumption aggregates from the Income and Expenditure Surveys of South Africa (IES) suggest that over the 5-6 year-period between 1994 and 2000, consumption growth has slowed to less than 1% per capita per annum, overall poverty headcount has remained unchanged, and the poverty gap has escalated (DLA/DoA, 2005). At the presentation of the 2004 Budget, the South African Finance Minister, Mr. Trevor Manuel, bemoaned the emergence of a '...*second economy* characterised by poverty, inadequate shelter, uncertain incomes and the despair of joblessness...' in which many South Africans are currently 'trapped' (Government of South Africa, 2004). According to Pauw (2005), a vicious circle of poverty is clearly evident and fuelled by the extreme disparities that put the greater proportion of national wealth in the hands of a small minority.

The former 'independent homeland' areas, namely Transkei, Ciskei, Venda, and Bophuthatswana, which were granted 'independence' by the apartheid regime (Berry, 1996; Raeside, 2004), have exhibited these problems much more than any other part of the country. According to Van Zyl and Binswanger (1996), these former 'homelands' were characterised by inadequate market access, poor and deteriorating infrastructure and support services for smallholder farmers. It is crucial for both national and regional development that the food marketing system is functioning optimally (Van Tilburg, 2003). As the National Department of Agriculture has observed, access to markets is imperative as a mechanism for integrating new and emerging farmers into the country's agricultural economy (DoA, 2000). This view is shared by several experts, including Jayne *et al.* (1997); Magingxa (2006); Rodrik (1998); Roe (2003); Summers (1992); UNDP (2003) and World Bank (2002, 2003). The Government of South Africa has therefore embarked on agricultural marketing reforms alongside a wide range of rural development programmes.

However, to date, these measures have produced little or no improvement in the circumstances of the rural smallholder producers whose conditions have either stagnated or actually become worse. Many researchers concluded that, for all practical purposes, the measures introduced to liberalise the domestic food market and integrate the country into the international system may actually have hurt rather than helped the smallholder farmers within the former homelands of South Africa (Makhura and Mokoena, 2003; Mokoena, 2002; Ndirangu, 2002; Nyamande-Pitso, 2001; Van Schalkwyk *et al.*, 2003). Why this is the case is not clear and provides grounds for more in-depth investigation. Without doubt, the phenomenal success of both the macro-economy and the commercial agricultural sector has largely by-passed the smallholder sector (Pauw, 2005; Perret, 2002; Pote, 2008). Given that these were the victims of the discriminatory policies implemented under the apartheid regime, there are grounds for concern.

Against this background, the present chapter aims to examine the extent to which the smallholder's predicament can be linked to constraints at the technical level, which mediate access to profitable marketing opportunities, among other difficulties, as well as determine by what mechanisms such insufficient market access hinders smallholder development.

4.2 The study area and research methodology

The study area for this study and the methodology adopted to collect and analyse the data are described in this section.

4.2.1 Study area

The geographic location of the Kat River Valley is defined by two major mountain ranges, namely the Winterberg and the Amatole Mountains around whose foothills it is situated. The Kat River Valley forms part of the Nkonkobe Local Municipality, which falls under Amatole District Municipality, and is divided into three different sub regions (see Figure 3.2 in Chapter 3), namely: the Upper Kat, Middle Kat and the Lower Kat. Prior to the end of the Apartheid regime in South Africa in 1994, the Upper Kat was part of the Ciskei 'independent' homeland. In terms of size, the Kat River Valley is approximately 80 kilometres in length and 1,700 km^2 in area, with the three main towns of Seymour, Balfour, Fort Beaufort and other smaller rural communities falling within its zone of influence.

Nel and Davies (1999) explain that racial conflicts in the early days of white settlement and ensuing apartheid regime resulted in laws that limited the access of the black population to the means of production, typically agricultural land. This situation has not changed much today (Aliber, 2005). Where the black population has moved out of the traditional villages into the towns, they are more commonly found in new settlements on the fringes where low-cost, poorly-designed housing are being constructed by the new government under the Reconstruction and Development Programme (RDP) (UN Habitat, 2001). These new settlements are the so-called 'locations' where facilities remain basic and poverty is rife, and all the features of underdevelopment are evident.

Bearing the foregoing in mind, the farmer surveys focused largely on the residents of these fringe settlements or 'locations' of the three towns, namely Fort Beaufort, Seymour and Balfour, which had been drawn randomly from the six main towns (the other three being Alice, Middledrift, and Hogsback) of the municipality. The farmers had previously been identified in the list received from the Chairperson of the Water Users' Association (WUA). In addition, other persons who were not members of the WUA but known to be actively farming were included in the sample.

4.2.2 Research methodology

A total of 80 farming households were drawn to represent diverse backgrounds of gender, education, and family situation, and enterprise specialisation such as arable farming and livestock production. General information on the institutional set up was obtained by open-ended interviews of community leaders and focus groups to complement information obtained through literature study. Subsequently, a single-visit household survey (n=80) was undertaken using structured questionnaires which covered a wide range of issues, including demographic information, costs and returns, marketing arrangements, and access within a broad definition. Given the objective of the study to identify the key technical constraints to market access, it was necessary first to come to an agreement on what measure of market access (or lack thereof) would be appropriate.

Ideally, a good measure of the technical constraints to market access for an agricultural entity should be the proportion of the physical output actually sold. However, the smallholder environment is characterised by shortcomings in educational attainment, which manifest in numeracy difficulties with their attendant measurement problems. In such a situation, estimates of production based on farmers' recall can hardly be reliable. For purposes of this particular study, it is only required to determine whether or not the farmer is able to market the farm produce, regardless of how much is produced in the first place. It would have been helpful to determine the proportion of physical output marketed as a way of determining the degree of market access enjoyed by producers. However, in the absence of reliable information on the amount of physical output produced, this is impracticable. It was therefore decided to merely determine whether all output intended for the market was sold as a measure of the constraint these farmers face in accessing markets.

The binary choice model was employed to estimate the probability that a smallholder farmer in the enumerated localities would have unsold produce in the farming season preceding the survey year. In this case, the basis for the analysis would be the reported market performance for 2006. Since only two options are available, namely 'all produce sold' or 'not all produce sold', a binary model is set up which defines Y=1 for situations where the farmer sold all produce, and Y=0 for situations where some or all produce was not sold. Assuming that x is a vector of explanatory variables and ρ is the probability that Y=1, two probabilistic relationships can be considered as follows:

$$\rho\,(Y=1) = \frac{e^{\beta' \chi}}{1+e^{\beta' \chi}} \qquad (1)$$

$$\rho\,(Y=0) = 1 - \frac{e^{\beta' \chi}}{1+e^{\beta' \chi}} = \frac{1}{1+e^{\beta' \chi}} \qquad (2)$$

Since Equation 2 is the lower response level, that is, the probability that some or all farm produce would not be sold, this will be the probability to be modelled by the logistic procedure by convention. Both equations present the outcome of the logit transformation of the odds ratios, which can alternatively be represented as:

$$\text{logit}\,[\theta(x)] = \log\left[\frac{\theta(x)}{1-\theta(x)}\right] = \alpha + \beta_1\chi_1 + \beta_2\chi_2 + \beta_3\chi_3 + \dots + \beta_i\chi_i \qquad (3)$$

thus allowing its estimation as a linear model for which the following definitions apply:
θ = logit transformation of the odds ratio;
α = the intercept term of the model;
β = the regression coefficient or slope of the individual predictor (or explanatory) variables modelled;
χ_i = the explanatory or predictor variables.

In line with Agresti (1990), Gujarati (1992), Harrell (2001), Hosmer and Lemeshow (1989) and Liao (1994), the right-hand term in Equation 3 is the natural logarithm of the modelled variables. A goodness-of-fit test following Hosmer-Lemeshow was conducted by examining the Pearson Chi-square outcomes calculated from the table of observed and expected frequencies as follows:

$$X^2_{HL} = \sum_{i=1}^{g} \frac{(O_i - N_i\pi_i)^2}{N_i\pi_i(1-\pi_i)} \qquad (4)$$

where:
N_i = the total frequency of the items in the ith group;
O_i = the total frequency of obtaining particular event outcomes in the ith group;
π_i = the average estimate of the probability that a particular event outcome in the ith group would be realised.

The goodness-of-fit test examined the displayed results for the Pearson and Hosmer-Lemeshow methods. As at least one of these measures gave a high enough ρ-value to dispel doubts about the model fitting the data, it was not necessary to consider alternative estimation procedures.

4.3 The variables

The variables examined in the study are presented in Table 4.1. Previous research has shown that market access is strongly influenced by such factors as the physical conditions of the infrastructure, access to production and marketing equipment, and the way the marketing functions are regulated (IFAD, 2003; Killick *et al.*, 2000). The variables are described below.

Location. The study was conducted in three towns. In order to accommodate this within the model structure and bearing in mind that a constant has been included, the location variable was included as two dummies for residence in Fort Beaufort (1) or otherwise (0), and residence in Seymour (1) or otherwise (0).

Age. This variable is expressed as the actual age of the household head in years. Previous studies, including Bembridge (1984), have established that this variable is a key determinant of behavioural patterns of household and community members. Younger farmers are

Table 4.1. Model variables applied in the analyses.

Variables	Unit	Type of variable	Expected sign (+/-)
Location	town in the municipality	categorical	
Age	actual in years	continuous	+/-
Household size	actual number	continuous	+/-
Educational level	attended formal schooling or not	categorical	+
Farming experience	actual years in farming	continuous	+
Access to credit	had or did not have access	categorical	+
Production loan	received or did not	categorical	+
NAFU membership	membership of farmer's union	categorical	+
Attendance at agricultural workshop	attended or did not attend	categorical	+
Non-farm income	had or did not have	categorical	+
Extension assistance	whether or not received	categorical	+
Extension visit	whether or not visited	categorical	+
Market distance	actual distance travelled	continuous	-
Total assets	actual value in rands	continuous	+
Crop income	actual value in rands	continuous	+
Livestock income	actual value in rands	continuous	+
Land size	actual size in hectares	continuous	+/-
Fertiliser use	whether used or not	categorical	+/-

expected to be more technically constrained than older farmers who are perceived to have acquired experience of farming and resources. Therefore, it is hypothesised that a higher age is negatively related to market access. This is supported by an observation by Mushunje *et al.* (2003) that older farmers are likely to have more resources at their disposal, which may make them more likely to cover costs of marketing more readily than younger farmers, despite being less aggressive to seek out more profitable markets. In that case, age may be related to the measure of market access either positively or negatively.

Household size. Increase in household size might increase the dependency ratio, which in turn affects savings and investment. Conversely, a larger household may mean increased labour availability, which enhances farm production under the kind of labour-intensive farming systems that prevail in communal agriculture. In turn, increased production increases the chances of market access due to larger economies of scale. Therefore, it is possible for either positive or negative relationships to exist between market access and household size.

Education level. Studies conducted in several developing countries have confirmed the importance of education in the decision-making process with implications for the socio-economic development and human capital production (Bembridge, 1984; Mushunje, 2005; Schultz, 1964). For the agricultural sector, earlier studies equally established that education plays an important role in the adoption or otherwise of improved practices in traditional agriculture (Bembridge, 1984). The absence of education is therefore expected to have a negative influence on these processes. In the light of that, it can be hypothesised that there is a positive correlation between education and market access.

Farming experience. This variable measures the number of years a farmer has been engaged in farming. It can be hypothesised that the lesser the number of years the farmer is involved in farming, the higher the probability of being technically constrained because certain farming techniques require that the farmer possesses some degree of experience. Thus, there is a positive correlation between market access and farming experience.

Non-farm income. This variable measures whether the farmer is receiving off-farm income. Off-farm income can help diminish on-farm technical constraints since the farm has alternative capital inputs. Farmers who lack off-farm income are likely to be affected by finance-related technical constraints than those who have. This is also supported by Mashatola and Darroch (2003). Thus, it can be hypothesised that there is a positive correlation between off-farm income and market access.

Land size. This variable refers to the size of land in hectares. Increase in land size may enhance production if the land is effectively utilised. At the same time, land may be available but not being effectively utilised. Effective utilisation will entail application of appropriate farm practices that will lead to higher physical output than otherwise would be the case. In the absence of more direct means of assessing effectiveness, this can only be inferred from the results. Intuitively, one can expect higher output if there is effective utilisation of available land, and lower output otherwise. It is also reasonable to expect that the more physical output a farmer produces, the more surplus is marketed. Therefore, it is hypothesised that there is either a positive or a negative correlation between market access and land size.

Market distance. This variable measures the distance to the point of sale of the farm output, notably a market centre where buyers congregate. The greater the distance to the market, the higher the logistical problems in terms of the availability of transport facilities and transport costs. Farmers who are located at considerable distances to the point of sale are likely to lack market access if they do not possess the means to transport their produce. Further, lucrative markets may be located far away from the point of production. It can therefore be hypothesised that there is a negative correlation between market access and distance to the market.

Extension contact. This variable measures whether farmers are in contact with extension officers more than twice a month. Extension service is an important source of farming information and advice to smallholder farmers (Enki *et al.*, 2001). Thus, it can be hypothesised that market access and extension contact will move in the same direction, the more extension contact with the smallholder the better the market access. In this case, two separate variables were employed to measure this attribute, namely frequency of extension visits and extension assistance.

Total gross income in 2006. Gross value of annual farm production from crop and livestock is an indicator of the performance of the farm business and the extent of commercialisation. Low values signify lack of market access and *vice versa* because farm income is a reflection of the value of surplus production. Total gross income is derived by combining crop and livestock income (presented separately in Table 4.1).

Value of assets. Inadequate technical farm inputs, tools, implements, farm machinery, motorised and other transport equipment, household appliances, residential facilities represent serious constraints to the average farmer. Tools and farm machinery are vital aids to field production while motorised transport are needed by farm household for transporting farm produce to markets. Household appliances such as radio and television are vital sources of information about market opportunities and prices. Assets can serve as collateral for credit. It is therefore expected that asset ownership and market access will be positively correlated.

Access to loans and/or credit. This variable measures whether farmers had access to institutional finance for the facilitation of production. Foltz (2005) developed a model that links credit access with agricultural profitability and investment in Tunisia. The findings show that credit constraint negatively affects farm profitability. As Reardon *et al.* (1996) have noted, farm profitability depends on availability of markets. It can therefore be hypothesised that market access is positively correlated to access to production loans and/or credit.

Agricultural workshop attendance. In South Africa, as in other parts of the world, attendance at technical workshops provides an opportunity for mass information sharing about opportunities and production possibilities, among other goals. This variable therefore measures the extent to which a farmer is exposed to agricultural education and training. Thompson *et al.* (2008) noted that workshops play a crucial role in influencing farmers' beliefs and attitudes in farming. It is hypothesised that market access is positively correlated to workshop attendance.

NAFU membership. This variable indicates whether the farmer has registered membership of the National African Farmers Union (NAFU), which is the umbrella association to which emerging black farmers subscribe. The principal aim of NAFU is to assist previously disadvantaged resource-poor farmers through networking and advocacy as well as provision of support services and targeted capacity building (NAFU, 2005). Therefore, membership of the association is expected to be a means for gaining various forms of advantage and accessing opportunities for marketing, among other benefits. Non-membership of NAFU can therefore constitute a major constraint and is hypothesised to be positively correlated with market access.

Fertiliser use. A number of studies have established that fertiliser usage is positively related to productivity (Reardon *et al.,* 1996; Xu *et al.*, 2009). Conversely, a farm unit that is too constrained to afford adequate amounts of fertiliser will most probably experience lower productivity which will translate to lower physical output and ultimately less marketable surplus. The production of insufficient volumes of the produce can discourage efforts to seek out outlets for disposal of produce. At the same time, retail outlets that buy up surplus produce from small producers are less eager to enter into contracts with small-volume producers. It is therefore reasonable to expect a direct relationship between fertiliser use and market access within the strict definition employed in this study.

4.4 Results and discussion

The summary statistics of the variables comprising demographic and production/marketing data are presented in Tables 4.2 and 4.3. In terms of the demographic characteristics of the sample, the summary statistics show that the majority of the farmers were male, married and aged about 57 years on average, with the youngest farmers being about 27 years old while one farmer was as old as 91 years of age. Household size ranged from 2 to 21 but averaged about 6.7 persons. The sample suggests that the majority of the farmers had some education, mostly up to 7 years of primary school education although some did not have any education at all. A few of the sample farmers had post-secondary education. There was evidence that some farmers supplemented their income by undertaking non-farm activities. It was also clear from the data that the majority of the surveyed farm households had been in the farming business for some time, with length of experience of up to half a century, although the data picked up a few new entrants into the farming business.

Regarding the distribution of assets and income across the sample, the study reveals a pattern that closely mirrors the situation in respect to the overall population of South Africa. For one thing, the data demonstrate pronounced inequities in terms of gross earnings from both livestock and crop production and the ownership of tangible/valuable assets in the study area. Table 4.2 presents the picture for both income and assets, among other variables.

As Table 4.2 shows, the gross value of farm produce ranged from nothing at all to as much as R158,000 (equivalent to about US$ 17,000) in one year, and the market value of assets ranged from those with negligible valuable assets to those with as much as R240,000 (equivalent to about US$ 25,000) at current prices. These data were not analysed for purposes of estimating household incomes but merely as a basis for classifying the households into

Table 4.2. Summary statistics of demographic and socio-economic variables (n=80).

Variable	Minimum	Maximum	Mean	Std. deviation
Location Fort Beaufort	0	1	0.51	0.503
Location Seymour	0	1	0.15	0.359
Age	27	91	57.50	14.635
Household size	2	21	6.73	3.048
Education level	0	1	0.25	0.436
Farm experience	1	55	21.45	10.376
Access credit	0	1	0.63	0.753
Production loan	0	1	0.72	0.449
NAFU	0	1	0.03	0.157
AG wkshp	0	1	0.70	0.461
Non-farm	0	1	0.41	0.495
Extension assistance	0	1	0.65	0.480
Total assets	0	240,300	23,125.95	45,621.839
Market distance	0	300	18.89	36.026
Crop income (Rands)	0	157,575	8,199.9	24,790.3
Livestock income (Rands)	0	135,000	9,416.6	19,989.12
Fertiliser use	0	1	0.24	0.428

rough socio-economic categories because comprehensive household income survey was not conducted. However, they do show that about half of the survey households lacked the possibility to earn much more than R20 per day (or about US$ 2) based on their reported gross value of farm income. The very high standard deviations of both asset value and gross farm income variables further confirm the huge disparities in socio-economic status even within the smallholder class, suggesting that this is by no means a homogenous category. The results similarly reveal other areas of inequalities in the smallholder sector and the serious constraints that this segment of the population still faces (Table 4.2).

A correlation matrix of the variables is presented in Table 4.3. The table presents the Pearson correlation coefficients to measure the strengths of the linear association between successive variables. According to the results, the correlation coefficients fall between r=0.00 and r=-0.48, the bulk of these indicating very weak associations. For instance, out of the 153 correlation coefficients calculated, 45% were less than 0.10, while 31% were between r=0.10 and r=0.25. Although little evidence of exact statistical independence was detected (generally at the default 5% level of significance), the associations either among the various variables or between any of them and the response variable, 'unsold produce', it was clear that

Table 4.3. Correlation matrix of modelled variables.

Variable	LocF/B	LocSYM	AGE	Hhs	Edlev	Farmex	Accrdt	Prdloan	Nafu	Agwshp	Nonfarm	Extast	Extvisit	Mktdist	Totaset	Unsold	Fertuse
LocF/B	1.0000																
LocSYM	-0.4307	1.0000															
AGE	-0.1711	0.1324	1.0000														
Hhs	0.1344	-0.1121	0.2375	1.0000													
Edlev	0.0433	0.0808	-0.0060	-0.1954	1.0000												
Farmex	-0.2000	0.1344	0.1345	0.1152	-0.2212	1.0000											
Accrdt	-0.2550	0.0702	0.0598	0.0097	-0.2509	0.2407	1.0000										
Prdloan	-0.0966	-0.0549	-0.1963	-0.0005	-0.1616	-0.0519	0.0655	1.0000									
Nafu	0.1562	-0.0673	-0.0110	0.3317	0.2774	0.0008	-0.0268	-0.0807	1.0000								
Agwshp	0.1255	0.1986	-0.0413	0.0396	0.1890	-0.1011	-0.3282	0.0855	0.1048	1.0000							
Nonfarm	0.0044	-0.2098	0.0236	-0.1251	-0.0733	0.2072	0.1485	-0.2801	-0.1342	-0.2272	1.0000						
Extast	0.1232	0.3083	-0.0541	-0.1185	0.0000	-0.0747	0.0526	-0.0411	-0.0504	0.3203	-0.1837	1.0000					
Extvisit	-0.0003	-0.2310	0.1034	0.1780	-0.0526	0.1066	-0.0491	-0.1525	0.1281	-0.2364	0.1046	-0.8415	1.0000				
Mktdist	-0.0499	-0.1512	0.0310	0.0999	-0.0337	-0.1287	0.1389	0.0880	-0.0062	0.0673	0.1104	-0.0426	0.0830	1.0000			
Totaset	0.3370	-0.1308	-0.0135	0.0581	0.1570	-0.0257	-0.1210	-0.3319	-0.0459	0.1088	0.0466	0.0306	0.0015	-0.0808	1.0000		
Unsold	0.3472	-0.1496	0.0068	-0.0199	-0.0228	-0.0462	0.0198	-0.0465	-0.0570	0.0604	0.1035	-0.0705	0.1406	-0.0342	0.3199	1.0000	
Fertuse	-0.1021	0.0123	-0.0293	0.0216	-0.0509	-0.0699	-0.1129	-0.3141	-0.0894	-0.0192	0.1290	-0.0831	-0.0175	0.0781	0.0308	0.0802	1.0000

the variables are sufficiently independent to be modelled together without any possibilities of multi-collinearity.

For the logistic regression analysis, the dataset was reduced from the 18 variables presented in Table 4.2 above, to 16 variables by dropping the crop and livestock income data to avoid overlap with the much more inclusive total asset variable. In addition, land size was deleted because, with land reform, all respondents could access land if they had the means. What would have been useful for analytical purposes was whether respondents had title to land, but since none had title; it was not helpful to model this variable. The first regression was run on SPSS, which accommodated all the variables and provided the results presented in Table 4.4. However, the very low Wald values suggest that location, farming experience, and membership of the black farmers' association were probably not very important. When regression was run on Stata-10, the variables that had low Wald values on SPSS were automatically dropped from the model and the results are presented in Table 4.5. The results from these two tables suggest that market access, represented by whether smallholders sold all surplus produce, was significantly related to credit access or receipt of production loan, the household's total assets, extension visits, and fertiliser use

Table 4.4. Results of logistic regression analysis (n=80).

Variables	Coefficient	Std. error of coeff.	Wald	Sig.[1]
Constant	-26.555	6,695.822	0.000	0.997
Loc F/B	20.615	6,695.821	0.000	0.998
Loc SYM	-0.892	11,853.988	0.000	1.000
Age	-0.011	0.041	0.077	0.781
Household size	-0.120	0.203	0.352	0.553
Education level	-0.227	1.394	0.026	0.871
Farming experience	0.001	0.048	0.001	0.976
Credit access	2.010	1.156	3.023	0.082*
Prod loan access	1.892	1.515	1.561	0.212
NAFU membership	-20.969	24,347.082	0.000	0.999
Workshop attendance	2.124	1.699	1.564	0.211
Extension assistance	-0.491	1.113	0.195	0.659
Extension visits	2.213	1.591	1.934	0.164
Non-farm income	0.548	0.966	0.322	0.570
Market distance	-1.054	1.280	0.679	0.410
Total asset	0.000	0.000	3.269	0.071*
Fertiliser use	4.48685	2.53085	1.77	0.076*

[1] * 10% significance level ($P<0.10$).

Table 4.5. Results of alternative logistic regression analysis using Stata 10 (n=80).

Variable	Coefficient	Std. error of coeff.	z	P>|z| [1]
Constant	-16.70562	8.296526	-2.01	0.044*
Age	0.0498933	0.0553111	0.90	0.367
Household size	-0.2120418	0.2768593	-0.77	0.444
Education level	1.285535	1.843297	0.70	0.486
Farming experience	0.0127572	0.0617426	0.21	0.836
Credit access	0.5860156	1.382694	0.42	0.672
Prod loan access	4.245034	2.304402	1.84	0.065*
Workshop attendance	-0.5247985	2.040503	-0.26	0.797
Extension assistance	4.805977	3.182111	1.51	0.131
Extension visits	2.925462	1.586681	1.84	0.065*
Non-farm income	1.616907	1.404886	1.15	0.250
Market distance	-0.0459667	0.0528308	-0.87	0.384
Total asset	0.0000241	0.0000116	2.08	0.038*
Fertiliser use	5.398449	2.86622	1.88	0.060*

[1] *stands for 10% significance ($P<0.10$).

Market access, represented by whether smallholders sold all surplus produce, appeared not to be significantly related to location, age of household head, household size, educational level, farming experience, National African Farmers' Union (NAFU) membership, attendance at agricultural training workshops, participation in non-farm employment, availability of technical extension assistance, hosting of extension advisory visit, and market distance.

The result with respect to production loans is consistent with expectations, given the widespread concerns about high production costs in the area and the fact that the bulk of the farmers do not have much by way of assets. Similarly, alternative sources of income were limited and most farmers derived their livelihoods from the farm. As Table 4.2 shows, only 42% of the respondents reported non-farm incomes while the rest derived all their incomes from farming. Therefore, smallholder farmers would keenly seek production loans to boost their capital base in any farming season, and this is expected to have some impact on their farming activities. The farmers pointed out a number of sources of such production loans during 2006 and 2007, including various instruments under the rural development programmes of the national and provincial governments. The results of the analysis predict a positive relationship between receipt of production loans and market access, which means that farmers who received this boost to their resource base would be expected to produce sufficient output for the market. The present study did not attempt to identify these various sources of production loans but they are obviously interesting subjects for more in-depth research to gain better understanding of how they operate and why not all farmers who wished to do so received the loans.

The household asset position is obviously an important variable in determining the ease with which smallholder farmers access markets and in whether they have a surplus to market. From the questions posed to the farmers, the study obtained information on the range of durable items owned by the household, including farming equipment, as well as storage and motorised transportation facilities. The results predict a positive relationship, which is consistent with common sense and economic theory given that these items directly contribute to production. The farmers enumerated in this study presented a picture of severe stress in respect to asset ownership and resource availability and it was clear that these would constitute crucial constraints on production expansion when they are in short supply.

Fertiliser use proxies the adoption of improved technologies and practices in the farming system. The model prediction of positive statistical significance makes intuitive and economic theoretic sense. Small farmers often experience difficulty with acquiring these vital resources and this has usually proved a serious constraint to output expansion. Although the fertility status of the soils in the research area was not assessed directly in this study, farmers' responses suggested that the soils are relatively infertile and there is a great need for supplementation with organic and inorganic fertilisers and manures. Many farmers utilise the droppings from their domesticated animals for this purpose but even that source is not always guaranteed. The University of Fort Hare Experimental Farms recently started charging a fee for its animal droppings, possibly indicating rising demand for them as substitutes for higher-priced organic fertilisers.

The variables that showed statistical significance or presented Wald values higher than 1.5 in Tables 4.4 and 4.5 above were re-inserted in the model to repeat the regression with unsold produce as response variable. The results are shown in Table 4.6.

As Table 4.6 shows, only extension visits and total assets retained their statistical significance. This result points to some tentative conclusions. In the first place, it seems that access to agricultural technical information through either extension visit or assistance and asset ownership, are the most important variables in determining the extent to which smallholders access markets in the project area. This predicted role for agricultural technical information is understandable especially within a broad definition that embraces such important market variables as prices, supply and demand. Without a doubt, agricultural technical information is crucial to marketing performance as well as knowledge about agricultural production. Within an agricultural system in which the extension services are well organised and the personnel are on hand to disseminate vital information on production and marketing issues, this role can be quite effective in promoting market access as defined in this study. In recent times in the country, this subject has become topical as newly settled farmers, who have been recipients of significant support from the government, complain that they are not

Table 4.6. Logistic regression on key predictors (n=80).

Variables	Coefficient	Std. Error of Coeff.t	z	P>|z| [1]
Constant	-6.495508	2.219351	-2.93	0.003**
Credit access	0.68806	0.6140748	0.42	0.263
Prod loan access	1.295268	1.10088	1.84	0.239
Workshop attendance	0.8580074	1.034635	-0.26	0.407
Extension visits	0.7128674	0.3844854	1.84	0.064*
Total asset	0.0000215	8.09e-06	2.08	0.008**
Fertiliser use	1.210035	0.8854498	1.88	0.172

[1] * 10% level of significance (*P*<0.10); ** 5% level of significance (*P*<0.05).

able to transform their farming operations into viable enterprises due to lack of information about opportunities for profitable sales of produce. The representatives of NAFU (National African Farmers Union) declared this a binding constraint during a phone-in public affairs programme on the national radio, SAFM (SAFM, 2009). The government is now putting a lot of emphasis on post-settlement support in a renewed bid to fast-track the land reform programme, which many analysts criticise as being extremely slow and not producing improvements in livelihoods for smallholders.

Asset ownership is an equally powerful force to the extent that it determines the ability to command resources for both production and marketing. The very significant influence of this factor reflects the complex environment in which the smallholder operates and the interrelationships among the farm business and the rest of the farm household. A smallholder with limited assets may be less capable of generating sufficient surplus for the market, or where such surplus exists, may have difficulty delivering it to the market, especially in situations where infrastructure is deficient. The same may be true for income. Poorer farmers will be more constrained than better-off farmers if transaction costs for accessing markets are high as is the case where infrastructure is deficient or other institutions are poorly developed. This relates also to economies of scale. Due to lower commodity volumes because of their limited scale, the smallholders are likely to have higher average costs than the established large-scale commercial farmers. This means that the smallholder farmers are likely to be more constrained than the better-resourced, large-scale farmers are in delivering produce to the market regardless of the physical distance and condition of the infrastructure. Thus, while insufficient market access may lead to erosion of farmers' income and asset base, low incomes and weak asset base may be the causes of the smallholders not being able to access markets even when such markets are available.

Market distance appeared not to have much influence on market access. A possible explanation for this is that market distance is a reflection of the state of the road and transport infrastructures, access to which is probably governed by asset ownership. A well-resourced farmer who owns serviceable motorised transport facilities, would be better able to access markets regardless of their distances as well as the condition of the roads linking them to the farms. In that case, it is the ability to produce a marketable surplus that is important rather than how to get it to the point of sale. That ability is already captured by the variable of asset ownership. However, it is possible that for two smallholder farmers with marketable surplus, market distance can kick in as a determinant of the relative ease with which either of them would access markets. One can argue that if the farm is already so severely constrained, in terms of insufficient information flow and lack of assets, that production is limited, availability or otherwise of infrastructure would have little chance of influencing the extent of marketed surplus. The same would be true if there is marketable surplus available but information is lacking about opportunities to sell them profitably.

4.5 Model adequacy

Tests to establish Goodness-of-Fit, assess the model fit, and absence of heteroskedasticity, were carried out on the data. Table 4.7 presents the results for the Goodness-of-Fit tests and indicates a good fit of the model represented in Table 4.6 at least in respect to the Hosmer-Lemeshow approach for which the ρ-values was higher than the chosen probability level for the logistic regression modelling (ρ=0.05). This implies that the predicted values of the response variable are close enough to the values observed in the study and the conclusion can be drawn that there is no difference between them.

While the indication is that some weak heteroscedasticity exists, the overall model test failed to confirm statistical significance. Therefore, no correction was considered necessary while taking note that the standard errors, t-statistics, and F-statistics in the model must be interpreted with some caution.

Table 4.7. Goodness-of-Fit tests on model fitted with all variables (see Table 4.6).

Method	Chi-square	*P*-value
Pearson	118.28	0.0002
Hosmer-Lemeshow	12.71	0.1221

4.6 Conclusions and recommendations

Income disparities are still substantial, even within a group that ordinarily would be considered homogenous. Thus, what one observes within the smallholder community is a reflection of the gross inequalities that characterise the wider South African society. Since the study did not carry out a comprehensive assessment of all income components, the model did not include income data *per se*. However, asset ownership was sufficient for purposes of classifying the households. Despite considerable farming experience and a commitment to farming demonstrated by their long-term planning, the majority of smallholders are unable to access the credit they need and therefore cannot acquire and use modern inputs such as mineral fertilisers and mechanical technologies. Access to publicly-provided infrastructure, such as good road networks, water and electricity, is also limited and many farmers do not have access to the right kind and amount of information and technical support that would make a difference in their skills levels about the farming business that make for profitability.

The indication that access to information, reflected in more frequent visits by the extension services, and asset ownership, had the most chance of influencing the extent to which the smallholder could sell marketable surplus during the year, confirms expectations regarding these variables within a largely family-oriented farming system. The significant influence revealed for asset ownership does explain largely why small farmers could still be physically excluded from markets despite a well-developed physical infrastructure even within the former homeland areas where these amenities are generally of poorer quality than in the rest of the country. As one can easily confirm by direct observation, the area has a reasonably good supply of all-season roads (despite some maintenance shortfalls), but with limited assets, the farmers are unable to take full advantage of them. Poorer small farmers are more severely constrained than the better off farmers to haul produce to markets or to cover costs of storage or delayed payment for produce sold. Although the present study did not explicitly assess quality issues and the ease, or otherwise, of compliance with standards, there is no doubt that smallholder farmers with limited means will experience greater difficulty meeting the quality standards of consumers for the basic commodities and will therefore be unable to sell profitably even if they could access the markets. As is well-known, price premiums due to higher quality/grades can be quite significant and often outweigh associated costs of marketing and may therefore lead to sellers reaping unusual profits.

While the limited scope of the present study in terms of geographical coverage and sample size call for caution in generalising these results, they agree with the findings of recent studies and popular viewpoints about progress within the sector. This means that these results have important practical implications for policy, especially in respect to conclusions regarding access to information and issues related to the strengthening of smallholder production capacity as a precondition for enhanced livelihoods through market participation. Many of the problems that smallholders confront have to do with knowledge about how to produce, what to produce, and where to sell what has been produced. These issues can be easily addressed by furnishing the smallholder farmers with the requisite technical information in a manner that they can conveniently process and use.

As part of the agricultural restructuring programme undertaken in the early to mid-1990s following multi-party elections, the agricultural parastatals established in the former homelands were dismantled. This meant that the technical services that these parastatals provided to smallholders were also discontinued. This move has proved quite premature and hasty, as it did not recognise the fact that these farmers would confront considerable knowledge gap in dealing with a more liberal market environment that accompanied the reforms. The gap created by the dismantling of technical services to smallholders is now beginning to hurt. This is true not only for the former 'independent homelands' of South Africa, but also for the rest of the country and rural Africa where smallholders are again being by-passed by the second Green Revolution, a point that has been made quite strongly in recent years by the Alliance for a Green Revolution in Africa (AGRA, 2010).

Any recommendations to improve market access for smallholders in the country must therefore be framed around the developmental context within which agricultural support has been implemented in recent years. Without equivocation, it is clear that the time has come to reinstate these services and align the agricultural extension services to the needs of smallholders. The agricultural extension services in the country are accepting this challenge as discussions are now increasingly focusing on the need for broadening the support services extended to beneficiaries of the land reform programme. There is a realisation that ownership of land does not answer all the questions about how to produce, what to produce and where to sell at a profit. Post-settlement support is now being introduced in various ways although the long-term sustainability of some of the recent interventions such as the farm recapitalisation by the newly created Department of Rural Development and Land Reform (DRDLR) is debatable. At least one aspect of this programme which involves the identification and deployment of strategic partners to deal with the dual problems of absence of technical information and asset constraints, promises to be quite controversial.

But whatever is decided and implemented, there is no doubt that the situation of the small-scale and emerging farmers will be enhanced quite significantly by arrangements that simultaneously address their skills gap (lack of technical knowledge and information) while ensuring that asset constraints are minimised. The mentorship relationship, which is implied by the strategic partnership provision under the recapitalisation programme, will be one way to achieve this. But the question will remain how such a scheme will be implemented to avoid exploitation of the small farmer and that possible conflict of interests on the part of the strategic partner does not become a source of serious conflicts in future.

References

Agresti, A., 1990. Categorical data analysis. John Wiley and Sons, New York, NY, USA.

Aliber, M., 2005. Overcoming underdevelopment in South Africa's second economy. 2005 Development Report, United Nations Development Programme (UNDP) and Development Bank of Southern Africa (DBSA), Pretoria, South Africa.

Alliance for a Green Revolution in Africa (AGRA), 2010. Strategy for an African Green Revolution. AGRA, Nairobi, Kenya. Available at: www.agra-alliance.org.

Bembridge, T.J., 1984. A systems approach study of agricultural development problems in Transkei. PhD thesis, University of Stellenbosch, Bloemfontein, South Africa.

Berry, B., 1996. Divisions of South Africa 1910-1994. Available at: www.allstates-flags.com.

Department of Agriculture (DoA), 2001. The strategic plan for South African agriculture. Department of Agriculture, Government of South Africa, Pretoria, South Africa.

Department of Labour, 2003. Speech by the Minister of Labour at Gala Dinner, Cicira College, Umtata 16 August 2003. SA Department of Labour, Pretoria, South Africa.

Department of Land Affairs/Department of Agriculture (DLA/DoA), 2005. Land and agrarian reform in South Africa: an overview in preparation for the land summit, 27-31 July, 2005, Ministry of Agriculture and Land Affairs, Pretoria, South Africa.

Development Bank of Southern Africa (DBSA), 2005. Development report 2005 – overcoming underdevelopment in South Africa's second economy. DBSA, Human Sciences Research Council and UNDP, Johannesburg, South Africa.

Enki, M., K. Belay and L., Dadi, 2001. Determinants of adoption of physical soil conservation measures in central highlands of Ethiopia: the case of three districts of Shewa. Agrekon 40: 293-315.

Foltz, J.D., 2005. Credit market access and profitability in Tunisian agriculture. Agricultural Economics 30: 229-240.

Government of South Africa, 2004. Budget Speech by Minister of Finance Mr. Trevor Manuel. Department of Finance, Pretoria, South Africa.

Gujarati, D., 1992. Essentials of econometrics. McGraw-Hill, Boston, MA, USA.

Harrell, F.E., 2001. Regression modeling strategies. Springer-Verlag, New York, NY, USA.

Hosmer, D. and S. Lemeshow, 1989. Applied logistic regression. John Wiley and Sons, New York, NY, USA.

Human Sciences Research Council (HSRC), 1996. A profile of poverty, inequality and human development in South Africa. Human Sciences Research Council, Pretoria, South Africa.

International Fund for Agricultural Development (IFAD), 2003. promoting market access for the rural poor in order to achieve the Millenium Development Goals. Discussion Paper, International Fund for Agricultural Development, Rome, Italy.

Jayne, T.S., J.D. Shaffer, J.M. Staatz and T. Readorn, 1997. Improving the impact of market reform on agricultural productivity in Africa: how institutional design makes a difference. Michigan State University International Development Working Paper No. 66, Michigan State University, East Lansing, MI, USA.

Killick, T., J. Kydd and C. Poulton, 2000. agricultural liberalization, commercialization and the market access problem in the rural poor and the wider economy: the problem of market access. IFAD, Rome, Italy.

Klasen, S., 1997. Poverty, inequality and deprivation in South Africa: an analysis of the 1993 SALDRU survey. Social Indicators Research 41: 51-94.

Klasen, S. and I. Woolard, 2005. Surviving unemployment without state support: unemployment and household formation in South Africa. Working Paper No. 129, University of Cape Town, Centre of Social Science Research (CSSR), Cape Town, South Africa.

Lam, D., 1999. Generating extreme inequality: schooling, earnings, and intergenerational transmission of human capital in South Africa. Research Paper, University of Michigan, Ann Arbor, MI, USA.

Liao, T.F., 1994. Interpreting probability models: logit, probit, and other generalized linear models. Sage Publications, Thousand Oaks, CA, USA.

Magingxa, L.L., 2006. Smallholder irrigators and the role of markets: a new institutional approach. PhD Thesis, Department of Agricultural Economics, University of the Free State, Bloemfontein, South Africa.

Makhura, M. and M. Mokoena, 2003. Market access for small-scale farmers in South Africa. In: L. Nieuwoudt and J. Groenewald (eds.) The challenges of change: agriculture, land and the South African economy. University of Natal Press, Natal, South Africa, pp. 137-148.

Mashatola, M.C. and M.A.G. Darroch, 2003. Factors affecting the loan status of sugarcane farmers using a graduated mortgage loan repayment scheme in KwaZulu-Natal. Agrekon 42: 353-365.

May, J., 1998. Poverty and inequality in South Africa. Report prepared for the office of the executive deputy president and the inter-ministerial committee for poverty and inequality. Praxis Publishing, Durban, South Africa.

Mokoena, R., 2003. Deterioration of agricultural support services. Report. National Treasury, Pretoria, South Africa.

Mushunje, A., A. Belete and G. Fraser, 2003. Technical efficiency of resettlement farmers of Zimbabwe. Contributed paper presented at the 41st Annual Conference of the Agricultural Economic Association of South Africa (AEASA), Pretoria, South Africa, 13 pp.

Mushunje, A., 2005. Farm efficiency and land reform in Zimbabwe. PhD thesis, University of Fort Hare, Alice, South Africa.

National African Farmers Union of South Africa (NAFU), 2005. NAFU-farmer. Avialable at: http://www.nafufarmer.co.za/pdf_files/NAFU_AD.pdf.

Nel, E.L. and J. Davies, 1999. Farming against the odds: an examination of the challenges facing farming and rural development in the Eastern Cape province of South Africa. Applied Geography 19: 253-274.

Ndirangu, N., 2002. Africa worse-off after agreement on agriculture. Paper presented at the Roundtable on Food and Trade: the WTO Development Change, November 4-5, 2002, Ottawa-Ontario, Canada.

Nyamande-Pitso, A., 2001. The experience of black business in the import/export market. Paper presented at the 8th Annual Agriculture Management Conference, October 30, 2001, Midrand-Johannesburg, South Africa.

Pauw, K., 2005. Quantifying the economic divide in South African agriculture: an income-side analysis. Provide Project Working Paper 3, 2005, Elsenburg, South Africa.

Perret, S., 2002. Water policies and smallholding irrigation schemes in South Africa: a history and new institutional challenges. Working Paper 2002 No. 19, University of Pretoria, Department of Agricultural Economics, Extension and Rural Development, Pretoria, South Africa.

Pote, P.P.T., 2008. Technical constraints to smallholder farming. MSc thesis, Department of Agricultural Economics and Extension, University of Fort Hare, Alice, South Africa.

Raeside, R., 2004. Flags of the world. Available at: www.allstates-flag.com/fotw/flags/za.

Reardon, T., V. Kelly, E. Crawford, T. Jayne, K. Savadogo and D. Clay, 1996. Determinants of farm productivity in Africa: a synthesis of four case studies. MSU International Development Paper No. 22. MSU, East Lansing, MI, USA.

Rodrik, D., 1998. Trade policy and economic performance in Sub-Saharan Africa. NBER Working Paper Series No. 6562, National Bureau of Economic Research, Cambridge, MA, USA.

Roe, T., 2003. Markets, trade and the role of institutions in African development. Paper presented at the pre-IAAE Conference, 13-14 August, 2003, at the President Hotel, Bloemfontein, South Africa.

South African Broadcasting Corporation-FM Radio, 2009. Phone-in public affairs programme. Johannesburg, South Africa.

Schultz, T.W., 1964. Transforming traditional agriculture. Yale University Press, New Haven, CT, USA.

Summers, L. H., 1992. Keynote address: knowledge for effective action. Proceedings of the World Bank Annual Conference on Development Economics 1991, World Bank Washington, DC, USA.

Thompson, G.K., J. Larsen and A. Vizard, 2008. Effectiveness of small workshops for improving farmers knowledge about ovine footrot. Australian Veterinary Journal 77: 318-321.

UN Habitat, 2001. Thematic Committee 6-8 June 2001: the South African housing policy: operationalizing the right to adequate housing. Available at: http://ww2.unhabitat.org/istanbul+5/1-southafrica.PDF.

United Nations Development Programme (UNDP), 2003. Human Development Report 2003 – Millennium Development Goals: a compact among nations to end human poverty. UNDP, New York, NY, USA.

United Nations Development Programme (UNDP), 2006. Human Development Report 2006 – Beyond scarcity: power, poverty, and global water crisis. UNDP, New York, NY, USA.

United Nations Development Programme (UNDP), 2007. Country programme outline for South Africa 2007-2010. UNDP/UNFPA, New York, NY, USA.

Van Schalkwyk, H.D., J. Groenewald and A. Jooste, 2003. Agricultural marketing in South Africa. In: L. Nieuwoudt and J. Groenewald (eds.) The challenges of change: agriculture, land and the South African economy. University of Natal Press, Natal, South Africa, pp. 119-136.

Van Tilburg, A., 2003. Framework to assess 'worldwide' the performance of food marketing systems. Paper presented at the Brown Bag Seminar of the Department of Agricultural Economics, November 11, 2003, Michigan State University, East Lansing, MI, USA.

Van Zyl, J. and H.P. Binswanger, 1996. Market-assisted rural land reform: how will it work? In: J. Van Zyl, J. Kirsten and H.P. Binswanger (eds.) Agricultural land reform in South Africa, Oxford University Press, Cape Town, South Africa, pp. 3-17.

World Bank, 2002. Empowerment and poverty reduction – a sourcebook. World Bank, Washington, DC, USA.

World Bank, 2003. World bank development report 2003 – Sustainable development in a dynamic: transforming institutions, growth, and quality of life. World Bank, Washington, DC, USA.

Xu, Z., Z. Guan, T.S. Jayne and R. Black, 2009. Factors influencing the profitability of fertilizer use on maize in Zambia. Agricultural Economics 40: 437-446.

5. Smallholders and livestock markets

Jan A. Groenewald and André Jooste

5.1 Historical context

Livestock marketing by smallholders in South Africa has to be seen against a background in which both history and tradition play important roles; these influences have had big influence on the keeping, utilisation and marketing of livestock by particularly black farmers, who constitute by far the larger proportion of smallholders in Sub-Saharan Africa, including South Africa.

Over a large part of the 18th and 19th centuries, expansion of the areas under control of first, the Dutch East India Company (DIC) which started the colonialisation of South Africa, and thereafter the British colonial government and also independent republics occurred not only by limiting the land available to the various Khoi, San and black tribes, but also by annexing land from these tribes. The loss of land led among some other results, to overpopulation of both humans and grazing animals in those areas, commonly known as the reserves (Grosskopf, 1930). At the end of the 1800s a long list of legislative measures restricting land ownership by blacks existed; one important motive appears to have been to ensure a continued supply of labour for the mines and for commercial agriculture (Kassier and Groenewald, 1992).

After 1910, a series of legislative measures increasingly restricted connection between black agriculturalists and commercial markets, the most important probably being the following (Kassier and Groenewald, 1992):

- The Land Act of 1913 prohibited 'natives' from buying or renting land outside the scheduled reserves without approval of the Governor-general (later the State President).
- The Native Trust and Land Act of 1936 made the Governor-General the trustee of all land tenure arrangements in the reserves.
- The Bantu Authorities Act of 1952 made the chiefs paid servants of the government.
- The Group Areas Act of 1956 divided the country into race areas and in essence forbade ownership of property across 'colour' lines.
- The Co-operative Societies Act of 1922 which was amended a few times, benefited white commercial agriculture, but these benefits did not cipher through to black farmers.
- The Marketing Act passed in 1937 and amended a few times later, provided support to commercial farmers, but not to subsistence farmers. All schemes under the act were administered and run by commodity boards in which producer representatives were majorities. Board members were not elected; but appointed by the Minister of Agriculture. This act was eventually repealed to be replaced by the must less drastic Agricultural Products Marketing Act in 1996.

By 1990, white farmers occupied 82.2 million ha of farm land (83.66% of the total) and black farmers occupied 16.1 million ha (16.34%).

5.2 Evolution of the beef sub-sector

Major legislative restrictions governed by the Marketing Act and implemented by the Meat Board were imposed on the South African red meat industry prior to deregulation in 1997. The following restrictions served as barriers to free market operations (Sunnyside Group, 1991):

- the distinction between controlled and uncontrolled areas (discontinued);
- restrictions on the creation of abattoirs (discontinued);
- the compulsory auctioning of carcasses according to grade and mass in controlled areas (discontinued);
- the compulsory use of agency services in controlled areas (discontinued);
- supply control via permits and quotas (discontinued);
- the setting of floor prices and the floor price removal scheme (discontinued);
- trade licensing and registration (still continued);
- controls on the sales of hides, skin and offal (discontinued);
- health and hygiene regulations (still continued);
- compulsory levies payable by producers (discontinued and reinstituted in 2005).

Most of these restrictions limited the producer's operation on a free enterprise basis, constrained the economic ability of traders, and limited the consumer's choice to what is supplied rather than to what is demand linked. Some of these restrictions are discussed in more detail in the next sections.

5.2.1 Distinction between controlled and uncontrolled areas

In order to prevent prices from falling below a desired level, the Meat Board divided the country into two area types: the main urban areas were denoted as controlled areas, while the rural areas were not controlled. Most consumption occurred in controlled areas, while most production occurred in non-controlled areas, including all the extensive livestock ranching areas.

Practically all meat marketed inside controlled areas had to be slaughtered within the controlled areas; transport of carcasses or meat cuts from the production areas was forbidden. Large abattoirs were developed for this purpose and with one exception, these abattoirs all belonged to one large parastatal company, called Abacor. Abacor thus had a virtually complete monopoly in urban livestock slaughtering and carcass delivery to the trade. Proponents of the system often argued that there were large economies of size in livestock slaughtering and in Abacor's operations, and that this was sufficient to offset the higher costs incurred by the system to transport live slaughter stock, rather than carcasses or meat cuts. In extensive research on this matter, Eales (1979) cited literature concerning research on abattoir scale economies in many countries, and noted that while costs per carcass or per kilogram of product decreased as abattoir sizes increased from very small to medium small, a stage is reached fairly early after which increased size did not further reduce per unit costs. Eales's own analysis confirmed that the South African urban abattoirs did not yield such scale or size benefits either; the only benefits were easier administration of the meat marketing scheme on the part of the Meat Board than would be the case otherwise. Eales (1979) went further and applied mathematical programming to determine an optimum pattern for abattoir location and operation in South Africa, given the situation of livestock production areas and transport networks. His results indicated medium-sized abattoirs spread over various parts of the country. It is interesting to note that within one decade after the repeal of the act and the dissolution of the Meat Board with its impressive structure, Abacor disappeared from the scene. It could not without the scheme's protection compete with more modestly sized operations within or closer to animal production areas.

5.2.2 Floor price system and quotas/permits

Floor prices were set for beef, sheep meat, goat meat and pigs meat carcasses in the controlled areas. The carcass sales were done by auction. Carcasses which did not reach the floor prices at the auctions were bought by the Meat Board, put into cold storage, and these carcasses were later, at what the Board deemed a good time, resold – usually as frozen carcasses.

The floor prices for meats were set at levels high enough to cause perennial surpluses and thus, mounting storage costs. In order to reduce such damages, the Meat Board embarked on two types of quota programs (see Groenewald, 2000): quotas were originally allocated to livestock agents who would then decide for themselves how much they should buy and from whom. After many complaints, the system was changed to one in which livestock producers could get marketing permits (permit being only semantically different from quota). The 'permit' and 'quota'systems co-existed. A permit or quota specified how many animals a producer or trader was allowed to deliver to a certain abattoir within a specified period of time.

Nieuwoudt (1985) showed that the permit/quota scheme increased consumer prices and prices at the main abattoirs and depressed the price farmers received on country auctions. The permit system aggravated the situation during periods of drought with the concomitant increased pressure to sell and increased the quota value[5]; the quota became more restrictive in limiting off-take from the veld and thus aggravated the farmer's position during adverse times. Those who could lobby most successfully obtained the majority of permits during

[5] During adverse time farmers were forced to sell animals, but if they did not have a quota they received very low prices. Thus, the demand for permits increased, resulting in increased quota values.

'over supply' periods while those whose voices did not count for much were left to sell their stock for the best they could get.

This obviously discriminated against producers who would not be able to do so – mostly stock grazers and smaller operators. Commercial producers in grazing areas and also smaller commercial producers complained that the system prevented them from marketing livestock in the season in which they would obtain the best prices; they were of the opinion that the system favoured the feedlots, big farmers and those close to the market over other livestock owners (Elliott *et al.,* 1987). Another study also found that feedlots and larger producers received preferential treatment (Lubbe, 1992). Under this regime, the feedlots became concentrated, with the five largest feedlots having a market share of 64%. The two largest feedlots belonged to two of the dominant firms in the meat trade (34% of capacity) (Lubbe, 1992).

It appears that under the Marketing Act, most commercial livestock holders suffered under discrimination on the main markets. The position was obviously much worse for stockholders in the black areas, or 'reserves', or 'homelands' as they were also called. Their only possible contact to the metropolitan markets could be through speculators who bought animals in their areas. In a sense these speculators also became monopsonists in these areas.

Significant policy changes took place from 1995 to 1997. In 1995 government started a process of liberalisation of the agricultural sector in line with its commitments made during the Uruguay Round of trade negotiations. In 1996 the new Marketing of Agricultural Products Act was legislated and implemented on 1 January 1997. The red meat industry started deregulation of the industry already in 1992, but the Meat Board was finally abolished in 1997. Overall the red meat industry made some significant adjustments since the 1970s, which guided the industry to become more mature to operate in the 'free' market system. These changes in strategic 'orientation' are highlighted by Ford (2006).

5.2.3 Opportunity driven

The 1970s was largely opportunity driven characterised low and inconsistent quality meat being offered to the market. Most of the trade in cattle was controlled by auctioneers (due to policy) and imports was restricted quantitatively. Grain farmers at the time saw the opportunity to market their grains alternatively and started feeding younger animals. At the time grainfed beef fetched 21% higher than normal market prices. The process was however hampered by primitive facilities (milling and handling) and outdated technologies and untrained staff.

5.2.4 Production driven

The gradual movement towards more intensive feeding of cattle introduced a time where the industry was largely production driven during the 1980s. This period saw stakeholders take cognizance of consumer priorities (although at a very low level). The industry expanded dramatically and a new grading system was introduced, but the Meat Board, through its regulatory functions, still limited optimal expansion. This period also saw the introduction of new technologies, the feedlot industry became more and more reliant on the grain industry for grain by-products and the industry was mainly controlled by three large companies that were vertically integrated. Growth promotants also became standard practice. By the end of the 1980s feedlots supplied 53.8% of beef to the commercial market.

5.2.5 Cost driven

During the 1990s the industry experienced both the effects of deregulation and liberalisation. This resulted in quantitative import restrictions being replaced by tariffs and high imports from especially the EU, which at the point was highly subsidised; imported beef attracted very high rebates. The deregulated and liberalised environment increased risk and resulted in lower prices; the outcome was that the 'three' big players in the market at the time disinvested in the red meat industry (largely because they were not able to adapt fast enough) leaving the opportunity for independent operators to grow. Moreover, the industry started feeling the pressures of a free market system and had to re-orientate itself to produce beef in a more cost efficient way. This period also saw the role of agencies (auctions) decline in importance and a movement towards direct buying and selling of animals.

5.2.6 Consumer driven

After the initial adjustments to a 'free' market system the industry increasingly realised the importance of the consumer as end user of the product at the turn of the 20th and 21st centuries. This led to greater emphasis in providing consistent quality, adding value, the development of standards and increased involvement further down the chain. The feedlot industry took the lead in this regard by amongst other things, buying out most of the main abattoirs, investing in deboning facilities and tanneries and investing in consumer surveys to better understand the needs and concerns of consumers. Moreover, the feedlot industry today in South Africa accounts for between 80 and 90% of beef marketed commercially. The evolvement of the feedlot industry was also stimulated by achieving better economies of scale in a 'free' market environment where profit margins are always under pressure.

5.3 Livestock keeping in tribal areas

By 2010, most smallholder farmers in South Africa are black people who have inherited traditions from their forbears. The great majority of these smallholders reside in former 'reserves' or 'homelands', where traditional land tenure systems still prevail and where traditional chiefs have power to influence farming practices and decisions. Reports of this power are possibly overstatements or exaggerations. For example, in a survey with a random sample of 350 small farmers in the erstwhile 'homeland' Lebowa – now part of Limpopo Province – the farmers were asked questions about decision-making concerning livestock. The decision about which livestock would be kept was in 72.6% of cases made by the husband, in 4.9% by the wife, in 1.8% by husband and wife and in 2.2% of cases by the chief or headman. Extension officers and stock inspectors were involved in the remaining 18.5% of cases. Decisions regarding selling livestock were in 80.9% of cases taken by the husband, 4.8% by the wife, 12.6% by husband and wife and only in 1.7% of cases by the chief or headman (Fenyes and Groenewald, 1985).

The literature, especially South African literature until the late 1980's, abounds with statements that in traditional African societies, livestock was of little economic significance and also statements that the livestock was mainly utilised for their hides, horns, meat, etc. and for ritualistic purposes (Monnig, 1969). Such statements are gross oversimplifications and indeed constitute misleading statements that indicate rather limited insight in economic logic.

Hughes (1972) distinguished between economic value and purely commercial value: cattle had historically been the only available source of readily transportable and convertible wealth for traditional farmers; there were absolutely no dependable banking and similar institutions in which assets could be safely deposited and withdrawn when the depositor desired to do so. This situation still preponderates, particularly in outlying parts of the former 'homelands'.

It has frequently been pointed out that livestock farming in Sub-Saharan African smallholder societies is very often characterised by overstocking, perverse supply response and low take-off from herds (Carlisle and Randag, 1970; Doran *et al.,* 1979; Lele, 1975). Explanations of this phenomenon often focus on cultural factors such as ignorance, traditional attitudes and traditional value standards.

Much has been written about the perverse supply response often encountered in smallholders' livestock marketing in different parts of Sub-Saharan Africa: increases in prices initially cause smallholders to increase their sales, but if prices rise above a certain level, they cut back on the quantity supplied – a case of the classical backward-bending supply curve. Some regression analyses (Doran *et al.,* 1979; Low, 1978; Low *et al.* 1980) tended to support this observation. However, observations done by Fenyes (1982) cast doubt on whether Lebowa smallholders reacted in this way.

This reaction has often been explained in cultural terms. It may however be more correct or realistic to explain it in terms of the distinction made by Hughes (1972) between economic and purely commercial criteria. In a more commercial environment with a plethora of financial institutions, a commercial producer will utilise rising prices to sell livestock. This will satisfy his cash needs and thereafter the money will be deposited in deposit-taking institutions (e.g. banks) and invested for further monetary gain. But what is the situation of a smallholder in a far away area without sound deposit-taking financial institutions? He will sell enough to satisfy his cash needs, and stop thereafter, as the livestock is his only available way to accumulate capital. He will be eager to sell with rising prices, but rising prices will sooner get him to the point of hoarding. He may even decide to expand his herd. One has to take into account that historically, he could not invest in land either.

Another reason why livestock farmers will sometimes sell more livestock in years characterised by lower prices than in years with higher prices, relates to weather variability. Drought years are characterised by deterioration of grazing and hence a short supply of feeding stuff for grazing animals. The condition of these animals deteriorates. Many farmers do then sell cattle to save them from starving and also to have more food per head available for the remaining animals. In such a year, the increased supply is also accompanied by falling prices. When more rain is experienced in ensuing years, livestock is held back, partially in order to replenish the breeding stock, and prices increase. This marketing behaviour is certainly not restricted to subsistence or smallholder farmers; it has also for long been a characteristic of commercial stock farmers (Louw *et al.,* 1979; Lubbe, 1992).

There are reasons to believe that stock owners in the areas under discussion have changed their motives for the keeping of livestock; commercial motives have become more important in Ciskei, presently part of the Eastern Cape Province (Fraser, 1992) and developments thereafter have certainly borne this out; popular farming magazines often have reports of smallholders entering the commercial livestock scene. It was also found that cattle owners in Bophuthatswana – now part of the North-West province – sold livestock because of money requirements (Groenewald and Du Toit, 1985). One factor that probably reduces the attractiveness of using livestock, particularly cattle, as a stock of wealth and hence encourages commercial selling is increasing levels of livestock theft as reported in the press.

The dearth of livestock and meat marketing opportunities and institutions in traditional areas such as the erstwhile South African 'homelands' has certainly contributed to meager marketing responses in such areas. Jooste and Van Rooyen (1996) argued that market access plays a pivotal role in the transition of the small-scale sector towards commercial production. Increasing market price variability (Jooste and Alemu, 2004), as well as a lack of infrastructure to support asset and income diversification (Bailey *et al.*, 1999), were identified as major causes of limited market participation. This will be illustrated by some examples in the next section of this chapter.

The number of animals in a person's possession also determines whether he/she will potentially become a commercial owner who wants to use marketing opportunities. Someone with less than 10 cattle, sheep or goats is unable to exploit these animals commercially as a meat producer. Reporting on a survey among cattle owners in four districts of Bophuthatswana, Groenewald and Du Toit (1982, 1985) reported that among owners with 10 or fewer cattle, 33.4% had sold cattle in the year of the survey. Of those with 11 to 20 cattle, the percentage sellers amounted to 52.9%, while 85.8% of those who owned more than 20 cattle also sold cattle. Fraser (1992) reported that in Ciskei, 80% of the small farmers who did not sell livestock gave insufficient numbers as the reason. The number of animals that would constitute an adequate quantity to participate in the market differed among the respondents.

5.4 Case studies

5.4.1 Lebowa

In a study in Lebowa – as previously mentioned, now part of Limpopo Province – Nkosi and Kirsten (1993) pointed out that quite a few authors had previously reported farmers in developing areas to have used a number of channels to market livestock: auctions, speculators, direct marketing, butchers and private sales. They found that in Lebowa, almost half of their respondents (48%) kept livestock for commercial purposes, while 45% kept it as a store of wealth. The reasons of the remaining 7% varied between ploughing, rituals, lobola (paying for brides) and prestige. Over the period 1982/1983 to 1991/1992, the distribution between the three main marketing channels was: auctions (42.23%), butchers (46.21%) and speculators (11.56%). At that time, auctioneering companies were appointed by the Lebowa Government. There were four auctioneering companies. Of the respondents who gave answers to a question of their satisfaction with the auctions, 64% were not satisfied. The two most often mentioned reasons for the dissatisfaction were low prices (42%) and not enough buyers (20%). The latter may indicate a lack of competition among buyers; this is probably the main reason for the low prices.

In an obvious imitation of the controlled marketing system which controlled meat marketing in the 'white' areas, speculators had to have proper permits issued by the Lebowa Marketing Board, and entry into speculation was an arduous process. Those with permits in reality became oligopsonists. The livestock owners were dissatisfied with the speculators (who marketed the animals obtained outside Lebowa, presumably in controlled markets); they complained that speculators were dishonest and treated the farmers badly. They evidently employed a range of tricks to convince farmers in need of cash to sell to them.

The farmers who sold directly to butchers were in general more satisfied with this marketing outlet. There do however appear to be size limits to this outlet.

The Lebowa experience indicates that governmental interference limiting numbers of buyers partaking in an outlet is a very unwise step or practice. It opens the way for abuse and is detrimental to the interests of small farmers who have to make a living from marketing products, including livestock.

5.4.2 Hammanskraal, Sterkspruit and Ganyesa

Hammanskraal, Sterkspruit and Ganyesa are three districts in separate provinces of South Africa: Limpopo, Eastern Cape and Northwest. Research on factors affecting market participation by small-scale farmers in these areas rather distant from each other has yielded striking similarities (Montshwe, 2006). The main findings were as follows:

1. Farmer training increases participation of small-scale farmers in mainstream livestock marketing. This could be expected, since marketing is facilitated by knowledge concerning product specification, price determination and timing. It was found that the great majority (87%) do not keep farm records, implying poor farm management skills. Access to market information – and the use thereof – also increases participation.
2. Larger herd sizes are associated with increased market participation rates.
3. Households who farm with livestock only participate more in mainstream livestock marketing than those with more diversified farming activities.
4. The results concerning distance from markets agree with results obtained elsewhere.
5. A somewhat surprising finding was that receipt of remittances has a positive impact on market participation. This may possibly be explained thereby that households use remittances to purchase production inputs which eventually translates into marketing surpluses.
6. Another surprising finding was that there is a significant, positive relationship between lobola and participation in mainstream cattle markets. This could be due thereto that lobola is nowadays not strictly paid in terms of cattle – cattle can be converted into cash.
7. The propensity to participate within the mainstream cattle markets increases with an increase in cattle mortality. This result suggests that since mortality is a source of risk, it will stimulate farmers to participate in mainstream markets in order to avoid further losses.
8. Theft of livestock has also, because it has become a source of risk, induced farmers to increase their participation in mainstream livestock markets. Selling of animals to avoid theft has been a phenomenon in many parts of South Africa; it has for example led to the liquidation of sheep herds in large parts of the Eastern Cape and Eastern Free State (CIAMD, 2002). The occurrence of theft also inhibits the ability of farmers to expand their herds to economically viable sizes.
9. As was also found in Kenya by Ouma *et al.* (2003) and in Botswana by Fidzani (1993), larger households are more inclined to participate in mainstream livestock markets.

Variables which did statistically not show any significant impact on market participation rates, include drought risk, extension visits, extension service, income level, birth in the family and land tenure.

5.4.3 Northern communal areas of Namibia

The northern communal areas constitute, like indeed the whole country of Namibia, an arid region and livestock rearing is the principal agricultural activity. The area is generally regarded as overgrazed and offtake in the form of livestock marketing will in the long run be a biological necessity. Farmers in this region are particularly disadvantaged in the marketing of meat or livestock because of a veterinary cordon fence – there to prevent the spread of foot and mouth disease – stretching from the west coast to the border with Botswana. This, together with the lack of local infrastructure, severely constrains meat and livestock marketing. Meatco, a parastatal marketing business, operates two modern abattoirs in the region. These abattoirs are far from the major livestock rearing districts (De Bruyn *et al.*, 2001; Duvel and Stephanus, 2000). Sartorius von Bach (1992) identified transport problems as the biggest weakness in the livestock marketing system.

Duvel and Stephanus (2000) studied stockowners' perceptions. Much of the other findings should probably be seen in the light of the farmers' purposes for keeping livestock. The purposes in order of ranking were as follows: cash for regular household support (72% of respondents), specific extraordinary purposes (66.2%), payment of lobola (46.6%), loaning cattle to others (41.8%), payment of tribal authority fines (48.5%), cattle slaughtering for ceremonial purposes (38%) and a much smaller percentage (16.8%) for commercial income generation. In declining order of importance, respondents rated stock diseases, lack of grazing (overgrazing), scarcity of water and droughts as the four most important stock farming problems. Poor markets ranked only seventh in this list, although as many of 58.2% of the respondents believed the marketing to be poor or very poor. Their first marketing preference was informal traders, followed by self slaughtering and selling, private sales and then Meatco. Compared with Meatco, respondents expected a 6% higher price from informal traders, 22% higher from private sales and 41% in the case of self-slaughtering and selling. Informal traders had a clear edge over Meatco in terms of accessibility (distance from selling point) and ease of selling.

In another study, De Bruyn *et al.* (2001) did a transaction cost analysis on meat marketing decisions in these tribal areas. They measured the effects of a as many as 30 variables that can influence transaction costs, and eventually summarised their main conclusions as follows:

1. Farmers with large herds or flocks are situated far from the Meatco buying points, but yet prefer selling to Meatco, even though they associate this marketing channel with high risk. Where owners have taken steps to mitigate this risk, sales to Meatco have increased. A point not mentioned in the authors' evaluation is that the rather limited size of alternative outlets may have been an important reason for the larger stock owners to prefer Meatco; Meatco is in a rather monopsonistic situation regarding sales to other regions.
2. Price information cost has had the biggest influence among the transaction cost variables – four times as large of the second most influential variable, price uncertainty.

3. Risk-associated costs of alternative marketing channels – particularly risk of extending credit, the availability of refrigeration and the opportunity costs of giving discounts to customers – have had a notable effect increasing sales to Meatco.

The authors (De Bruyn *et al.*, 2001) recommend that Meatco should move its buying operations closer to the districts with larger cattle herd owners and focus on them as a group. Information cost can be remedied, e.g. information circulation including radio broadcasts, extension officers and poster advertisements. Information should include data on expected prices as well as places and times of stock sales. Informal markets can also be upgraded, e.g. through refrigeration facilities.

The authors did not comment on the monopsonistic position of Meatco in the Namibian meat market.

5.4.4 Eastern Cape province

In the Eastern Cape Province, the nature of smallholder involvement in livestock marketing is, according to findings by Coetzee *et al.* (2005), in many cases rather similar to what has been reported in Lebowa, Northern Namibia and elsewhere. Overgrazing is a serious problem, and market participation is rather small – under 10% of the total herd, whereas according to Jooste (1996), the take-off in commercial farmers' herds ranges between 23% and 25%. Farmers' reasons for keeping livestock include the same type of human dimensions as argued to be the case, and found in other areas. Livestock is also marketed at informal markets, to speculators and at auctions. The informal markets are characterised by seasonality and poor market information pertaining to both prices and quality required.

In dealings with speculators, farmers often sell their livestock below market value due to poor information, bad timing and because they are in a weak bargaining position. Fidzani (1993) found similar trends in Botswana.

As far as auctions are concerned, the marketing boards instituted in homeland days by the Transkei and Ciskei governments ceased to exist in 1992. Isolated efforts to improve livestock marketing have since 1999 been initiated by the Government and donors. These programs included the Agrilink Project which initiated rural livestock auctions. It was successful as long as the Government supported it financially.

Coetzee *et al.* (2005) identified the following as the major marketing constraints faced by small-scale farmers:

1. *Poor condition of livestock*. A lack of buyers is frequently given as a major reason for small scale farmers' inability to access markets, but speculators and auctioneers raise concerns that they cannot pay competitive prices for livestock that is in a poor condition or not ready for the market. This leads to low prices, especially during dry spells. Neither can

old animals be expected to fetch good prices. Poor condition is often attributable to inadequate grazing and degradation of the natural resource.
2. *Lack of marketing information.* Asymmetric information reigns in the market; auctioneers and speculators normally are better informed of marketing conditions. This leads to frustrations during negotiations at auction sales and with private speculators. In its turn, low rates of price acceptance reduce the number of buyers willing to deal with the farmers.
3. *Problems with livestock identification.* The Livestock Identification Act renders livestock branding and marking compulsory. Many small farmers however do not brand or mark their animals, mostly for the following reasons (USAID, 2003): the cost of registering a unique brand and the cost of branding or marking equipment discourage many small farmers; stray animals on roads cause accidents, and owners want to avoid the possibility of claims against them; and a lack of facilities to brand or mark animals. This has caused many farmers to be turned away from auction sales.
4. *Lack of infrastructure.* This involves both physical (communication, transport and roads) and institutional (market information, security and animal disease control) infrastructure. As elsewhere, many small-scale farmers are in areas remote from market paces, and roads are in many parts in a very poor condition.
5. *Poor production and marketing management.* The small-scale farmers are short in these skills, and improvement herein is vitally necessary.

5.5 Strategies to improve livestock marketing[6]

At least seven facets ought to be included in strategies to improve livestock markets for small scale farmers.

5.5.1 Motivational aspects

It has been pointed out earlier that a lack of other investment opportunities – such as banking deposits for interest and other investments with capital growth opportunities – has probably been a major cause for the practice by many smallholders to keep particularly cattle, but presumably also other livestock, as a storage of wealth. This phenomenon has certainly contributed to overstocking and has proved in times of drought to be a rather weak way to prevent losses. The provisioning of deposit and investment opportunities together with sound advice in particular the off-lying areas is likely to present to smallholders opportunities that should both prove to be better storages of wealth and contribute to reduce overgrazing. The banking sector, and particularly micro-credit concerns (see Chapter 8) can in this respect play a determinate role.

[6] A large part of this section is based on Coetzee *et al.* (2005).

Other local social infrastructure – e.g. improved schools – is also needed to broaden the windows of opportunity in the areas in which most smallholders are concentrated. This should increase demand for cash and thus act as motivation to increase livestock turnover rates. Thus, the State should become involved – not in livestock marketing *per se,* but in creating conditions for prosperity. This will also increase local meat demand and the activities of local traders.

5.5.2 Improved quality and condition of livestock

The condition of the animals offered plays an important role to determine their marketability. Thus, concerted efforts should be made to improve grazing conditions. There is certainly a possibility of forming co-operative feedlots, run by livestock owners in many smallholder areas. Bophuthatswana farmers have quite some time ago expressed a willingness to market to feedlots (Groenewald and Du Toit, 1985). Farmers will however need guidance – and mentoring – in setting up the organisation for such feedlots, and in the practice of feedlotting. In such an environment, improved livestock management is an essential requirement for progress (Kumar *et al.,* 2000).

5.5.3 Market information

Without proper market information, farmers cannot hope to become successful entrepreneurs. This function should be fulfilled by extension services, the agricultural press and bearing in mind the remoteness of many areas as well as low prevailing levels of literacy, by private agencies involved in the meat trade. The problem of asymmetrical information should however be borne in mind, and strategies should be devised to overcome this.

5.5.4 Access to livestock identification facilities

It has been reported that in at least the Eastern Cape Province there are activities under way aimed to accelerate the identification process. This should obviously be done in all the livestock areas in every province. The main aims must be accessibility and cost reduction for small farmers. Cooperatives can potentially play an important role in this respect.

5.5.5 Marketing infrastructure

It is well known that in South Africa – as in the rest of the continent – rural roads are generally of poor quality and in poor condition. This has severely constrained local development, particularly of farming, including livestock production. This should be a much higher priority of national, provincial and local government than has thus far been the case. According to Kgantsi and Mokoene (1997) lack of properly maintained roads, telephones, fencing water and electricity make it very difficult for small farmers to run farming operations. BATAT (2004) has reported that in Government efforts to provide marketing infrastructure, they neglected the participation of communities, farmers and traders. This should be corrected.

It is also important to see to it that livestock sales venues are properly accessible to all potential participants. This does also tie in with proper road links as well as proper publicity to livestock marketing actions.

Serious attention ought be given to the possibility of establishing one-stop service centres where farmers can sell their livestock, assess financial services, obtain inputs such as livestock remedies, feed and supplements, sell hides and skins and attend training courses. Montshwe (2006) recommended it to be at municipality level or ward in the case of deep rural areas.

5.5.6 Crime prevention

Problems with stock theft have, as stated before, become very serious in certain parts of the country, including areas with concentrations of emerging or small farmers. It is imperative for the police services to improve their ability to combat this problem.

5.5.7 Farmer education

There is certainly a need to improve literacy and numeracy levels. This pertains to all active age groups. Reports in the press about the failure of the school system to provide sufficient literacy and numeracy to school attendees, is a matter of grave concern. Training programs should include visual aid materials and adequate illustration by weighing animals and applying current market information (prices per kilogram) to determine current market value. Farmers' negotiating skills should also receive attention. The fact that farm management skills in general are underdeveloped, also implies that training will have to accentuate – among animal producing skills – farm management, marketing and financial management. This may have to involve special training sessions for extension personnel.

5.5.8 Institutional support

Institutional support should be provided by the State, the private sector and farmers' organisations (both emerging and commercial farmer organisations) cooperating with one another with the common goal to improve the living, production and marketing conditions of smallholders.

Moreover, according to Jooste (2007), the application and design of the value chain to absorb this group of farmers will require additional steps and a proper delineation of responsibilities than what the case would have been if only the commercial sector was applicable. In this regard Public-Private Partnerships will be vitally important, since on the one hand the private sector has the core competencies and tacit knowledge, but not the resources to provide a comprehensive support service nor do they want to take the risk, and on the other hand government support services are severely constrained in terms of capacity and tacit knowledge but have the reach to service this group of farmers (Jooste, 2007). The model proposed by Jordaan and Jooste (2003) can be a useful tool in conjunction with the value chain approach to address the challenges this sub-sector face. Figure 5.1 illustrates the proposed model.

Figure 5.1 shows a holistic framework that would cater for the needs of all possible role-players concerned. Figure 5.1 is divided into three distinct levels, namely subsistence (level 1), emerging commercial farmers (level 2) and commercial farmers (level 3). At each level the degree of involvement by government and private sector role-players should be different. Figure 5.1 suggests that the sole responsibility for support on the subsistence level should reside with government. However, some subsistence farmers have the potential, and in fact develop into emerging commercial producers, that are depicted in the middle of Figure 5.1. Support to this group of farmers should be in the form of an alliance that includes government, private sector, academic institutions and commercial farmers' initiatives.

It is important to note that the provision of support by government should extend over both levels 1 and 2. The reason for government also extending its function to level 2 is that this group of producers is not ready at this stage to enter the commercial market insofar technology gathering and adoption, as well as management skills are concerned, yet they do not qualify as subsistence farmers targeted by international and governmental support programs. As they move towards complete commercialisation (level 3) the support functions

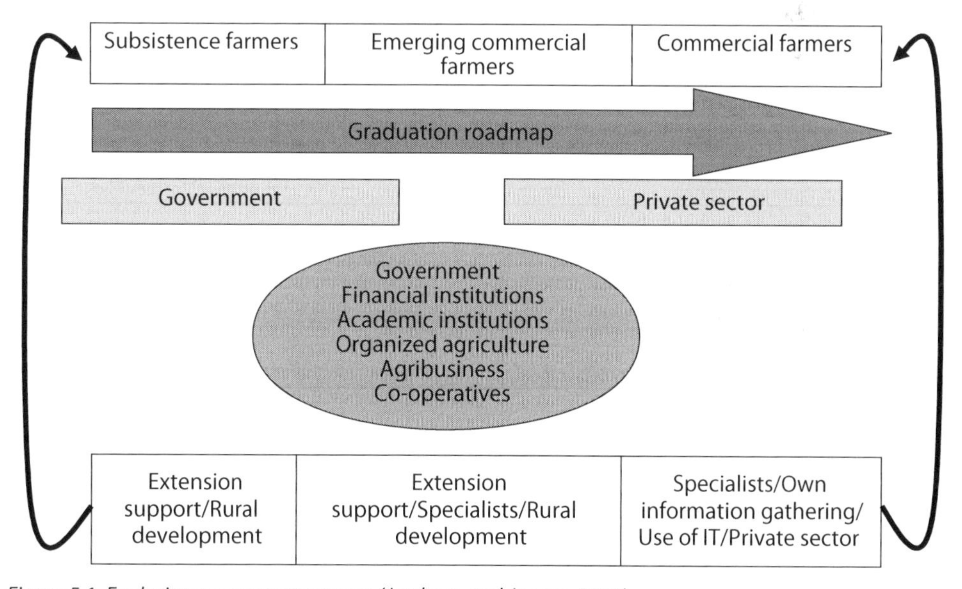

Figure 5.1. Evolutionary support process (Jordaan and Jooste, 2003).

performed by government could be transferred to the other role-players. This framework also entails a risk-bearing portfolio that is acceptable for all concerned, i.e. as farmers graduate towards commercialisation risks are reduced to levels that are acceptable for the private sector.

The proposed framework is necessary to close the gap between the commercial and non-commercial sector. Only when this gap is sufficiently closed will the emerging farmers be able to benefit from a market that is value orientated since closing the gap also translates into sharing of knowledge assets and lower risk for participants involved in this market (risk can be seen as product, price, human and financial risks).

This will improve the image of the state; the private sector will likewise benefit in the form of improved image and improved profit potentials.

5.5.9 Grazing rights and management

In communal areas, grazing rights and stocking rates are serious matters inhibiting the ability of stockowners to become prosperous and sustainable commercial producers. In communal grazing regimes, individual stockowners cannot through individual decision-making manage grazing so as to prevent overgrazing and to maintain stocking rates close to the long-run optimum. There is an inherent incentive for the individual to increase his stock numbers. This problem deserves much more serious attention than it has thus far received. It certainly warrants intensive research. Solutions of this problem will have to be found through interaction between economic, pasture, zoological, social and political scientists; positive cooperation with and between both tribal authorities and communities will be needed.

5.5.10 Feedlotting

Feedlotting is an activity which is rather seldom encountered in communal areas. It can however possibly offer some solution to some of the problems mentioned in this chapter. Feedlotting will have to be well organised if it is to succeed in communal areas. The goodwill of local leaders and indeed whole local communities will be prerequisite. The participants will in such a development have to produce at least part of the feed themselves, or obtain particularly the bulky foods from nearby producers. If they obtain this fodder from other smallholders – including small holders with very little or no livestock of their own, this may have multiplier economic effects within communities. Other prerequisites for success will be the organising of marketing outlets and transport; the poor state of road infrastructure in many tribal areas is likely to by a major hazard.

5.6 Conclusion

It can be concluded that there are severe weaknesses and challenges in smallholder livestock production and marketing in South Africa. It is of utmost importance for the future of the whole country that these weaknesses be overcome and the challenges faced. Failure to do so may very likely cause ecological, followed by economic disaster because of the long-run effects of severe overgrazing. This will need concerted effort by every organisation and every individual involved.

References

Bailey, D., C.B. Barrett, P.D. Little and F. Chabari, 1999. Livestock markets and risk management among East African pastoralists: a review and research agenda. Research report, Utah University, Logan, UT, USA. Available at: http://www.forum.mn/res_mat/res_mat-184.pdf.

Broadening Access to Agriculture Thrust (BATAT), 2004. Report of BATAT: marketing design team. Compiled by Jerry Tube. http://www.nda.agric.za/docs/BATAT/marketing.htm#1.%20 INTRODUCTION.

Carlisle, K.R.M. and A.G. Randag, 1970. Analysis of factors affecting meat packing developments. In: Bunting, A.H. (ed.) Change in agriculture. Croom Helm, London, UK, pp. 229-236.

Chair in International Agricultural marketing and Development (CIAMD), 2002. Livestock outlook. University of Free State, Chair in International Agricultural marketing and Development, Bloemfontein, South Africa.

Coetzee, L., B.D. Montshwe and A. Jooste, 2005. The marketing of livestock on communal lands in the Eastern Cape province: constraints, challenges and implications for the extension services. South African Journal of Agricultural Extension 34: 81-103.

De Bruyn, P., L.N. De Bruyn, N. Vink and J.F. Kirsten, 2001. How transaction costs influence cattle marketing: decisions in the northern communal areas of Namibia. Agrekon 40: 405-425.

Doran, M.H., A.R.C. Low and R.L. Kemp, 1979. Cattle as a store of wealth in Swaziland: implications for livestock development and overgrazing in Eastern and Southern Africa. American Journal of Agricultural Economics 61: 41-47.

Duvel, G.H. and A.L. Stephanus, 2000. A comparison of economic and cultural incentives in the marketing of livestock in some districts of the northern communal areas of Namibia. Agrekon 39: 656-664.

Eales, R., 1979. A long-term physical abattoir development plan for South Africa. PhD thesis, University of the Witwatersrand, Johannesburg, South Africa.

Elliott, M.B., W.L. Nieuwoudt and M.C. Lyne, 1987. An opinion survey on aspects of beef marketing in South Africa. Agrekon 26: 62-65.

Fenyes, T.I., 1982. A socio-economic analysis of smallholder agriculture in Lebowa. DSc(Agric.) thesis, University of Pretoria, Pretoria, South Africa.

Fenyes, T.I. and J.A. Groenewald, 1985. Food production in Lebowa: the interaction of social, physical and economic considerations. South African Journal of Agricultural Extension 14: 46-56.

Fidzani, N.H., 1993. Understanding cattle offtake rates in Botswana. PhD thesis, Boston University, Boston, MA, USA.

Ford, D., 2006. The South African feedlot industry. Presentation made during a Meat Imbizo at the University of the Free State, Bloemfontein, South Africa.

Fraser, G.C.G., 1992. Farmer response to the provision of livestock marketing facilities in Ciskei. Agrekon 31: 104-106.

Groenewald, J.A., 2000. The agricultural marketing act: a post-mortem. South African Journal of Economics 68: 364-402.

Groenewald, J.A. and J.P.F. Du Toit, 1982. Some features of livestock farming in Bophuthatswana. Development Studies Southern Africa 4: 225-242.

Groenewald, J.A. and J.P.F. Du Toit, 1985. Marketing behaviour and marketing preferences of Bophuthatswana cattle owners. Agrekon 24: 24-29.

Grosskopf, J.F.W., 1933. Die plek van die Bantoebevolking in die Suid-Afrikaanse volkshuishouding. South African Journal of Economics 1: 451-466.

Hughes, A.J.B., 1972. Land tenure, land rights and land communities on Swazi nation land in Swaziland: a discussion of some interrelationships between the traditional tenurial system and problems of agrarian development. Institute of Social Research, University of Natal, Pietermaritzburg, South Africa.

Jooste, A., 1996. Regional trade of beef in southern Africa. MSc (Agric.) thesis, University of Pretoria, Pretoria, South Africa.

Jooste, A., 2007. Evolution and drivers of the South African beef chain. Invited paper presented at the VI International PENSA Conference entitled 'Sustainable Agri-food and Bioenergy Chains/ Networks Economics and Management'. University of São Paulo, Ribeirão Preto, 24-26 October 2007, São Paulo, Brazil.

Jooste, A. and C.J. Van Rooyen, 1996. Access to emerging farmers in the red meat industry. Paper presented at the All Africa Conference on Animal Agriculture, Pretoria, South Africa.

Jooste, A. and Z.G. Alemu, 2004. The impact of the exchange rate on beef prices and imports. Working Paper, University of the Free State, Bloemfontein, South Africa.

Jordaan, A. and A. Jooste, 2003. Strategies for the support of successful land reform: a case study of Qwa Qwa emerging commercial farmers. South African Journal of Agricultural Extension 32: 1-14.

Kassier, W.E. and J.A. Groenewald, 1992. The agricultural economy of South Africa. In: C. Csaki, T.J. Dams, D. Metzger and J. Van Zyl (eds.) Agricultural restructuring in Southern Africa. Association of Agricultural Economists of Namibia, Windhoek, Namibia.

Kgantsi, M. and S. Mokoene, 1997. South African farmer support services: an end-user perspective. Unpublished Report, Development Bank of Southern Africa, Pretoria, South Africa.

Kumar, S., M. Candler and P.C. Harbola, 2000. Livestock-based farming system – a case study of Kumaon Hills. ENVIS Bulletin (Himalayan Ecology & Development) 8(2).

Lele, U., 1975. The design of rural development: lessons from Africa. John Hopkins University Press, Baltimore, MD, USA.

Louw, A., J.F.W. Grosskopf and J.A. Groenewald, 1979. Beef production systems and sales strategies in an extensive ranching region in South Africa. Agricultural Systems 4: 104-114.

Low, A.R.C., 1978. Cattle supply responses in Sudan and Swaziland: motivational references and overgrazing implications. Oxford Agrarian Studies 7: 62-74.

Low, A.R.C., R.L. Kemp and M.H. Doran, 1980. Cattle wealth and cash needs in Swaziland: price response and rural development. Journal of Agricultural Economics 31: 225-235.

Lubbe, W.F., 1992. The red meat marketing scheme: an evaluation in a dynamic environment PhD thesis, University of Pretoria, Pretoria, South Africa.

Monnig, H.O., 1967. The Pedi. Van Schaik, Pretoria, South Africa.

Montshwe, B.D., 2006. Factors affecting participation in mainstream cattle markets by small-scale cattle farmers in South Africa. MSc (Agric.) thesis, University of the Free State, Bloemfontein, South Africa.

Nieuwoudt, W.L., 1985. An economic analysis of demand and policies in the beef industry. Agrekon 34: 9-13.

Nkosi, S.A. and J.F. Kirsten, 1993. The marketing of livestock in South Africa's developing areas: a case study of the role of speculators, auctioneers, butchers and private buyers in Lebowa. Agrekon 32: 230-237.

Ouma, E.A., G.A. Obare and S.J. Staal, 2003. Cattle as assets: assessment of non-market benefits from cattle in smallholder Kenyan crop-livestock systems. International Conference of Agricultural Economists, Durban, South Africa.

Sartorius Von Bach, H.J., 1992 Supply response in the Namibian beef industry. MSc (Agric.) thesis, University of Pretoria, Pretoria, South Africa.

Sunnyside Group, 1991. The red meat industry: assessment and recommendations. Report September 1991. Sunnyside group.

USAID, 2003. Agrilink II Project 200, Monthly Progress Report # 22. USAID, Washington, DC, USA.

6. Unlocking markets to smallholder farmers: the potential role of contracting

Jan A. Groenewald, Jacobus Klopper and Herman D. van Schalkwyk

6.1 Introduction

> In every 'underdeveloped' country I know of, marketing is the most underdeveloped or least developed part of the economy....[Marketing] would make the producers capable of providing marketable products by providing them with standards, with quality demands and with specifications for their product. It would make the product capable of being brought to markets instead of perishing on the way (Drucker, 1958).

This quotation from one of the best-known management and business writers of the 20th century, underlines one of the crucial matters which have impeded particularly rural development in many parts of the world, including parts of South Africa. Without access to knowledge, modern inputs and well-directed outlets, the smallholder farmers who have grown up under conditions of poverty are doomed to keep on living in poverty. Many publications have revealed very low market participation rates of smallholders in the marketing of farm products. For example, only the following proportions of smallholders in South Africa's Limpopo province sold the following product types in 1997: horticulture (19%), livestock (17%), maize (21%), and other field crops (22%) (Makhura, 2001).

Referring to a number of publications, Delgado (1999) states that African smallholder agricultural development will need to overcome many structural constraints emanating from both history and geography. This includes a need for African smallholders to become increasingly involved in the production and marketing of goods that have high value relative to weight. This will require significant vertical integration of smallholders to processing and marketing firms (Delgado and Siamwalla, 1997).

The question is how to achieve this in a time of rapid change in the marketing of agricultural products and also in view of the competition of established (and more experienced) commercial producers.

6.2 Changes in agricultural marketing

Most countries in the world have since late in the 20th century undergone relaxation of government controls in agricultural marketing. In South Africa, the Marketing Act, first promulgated in 1937 and amended a few times later, that instituted very strict control over the marketing of most farm products, was repealed in 1996 and replaced by a new act, the Marketing of Agricultural Products Act. This act allows free trade in farm products, while maintaining the right to intervene in certain types of crisis situations. Monopolisation is controlled under monopoly legislation. Various investigations into monopoly behaviour in food chains and also farm input chains have been done since the late 1990's and heavy fines levied in cases.

Changes since the deregulation of agricultural marketing have been summarised as follows (Vojtech *et al.*, 2006): a freer pricing system, less reliance on state support, improved efficiency, a more competitive and less regulated food chain and sharp growth in agricultural exports. South African agriculture has thus increasingly become part of the globalised agricultural trade structure. This means that South African producers have to adapt to the requirements as set by consumers in the main importing countries, which in monetary terms mainly means the developed countries of the northern hemisphere.

These consumer requirements include quality requirements – related to among others, health standards, taste, appearance, some ethical considerations – and matters related to availability where and when consumers want the products. These requirements do, with a certain time lag, also spread to (particularly) better off customers in export countries such as South Africa. This leads to a situation in which different segments of the consuming public will tend to support different sections of the retail trade based on geographical location, product mixes and price.

Retailer groups have developed with large expansions of supermarket chains establishing dominance in retail marketing of most foods, and a similar phenomenon has occurred in food and beverage manufacturing. Vermeulen *et al.* (2008) cite the analysis of Fedderke and Szalontai (2005) that the South African food manufacturing sector is characterised by a high level and slightly increasing level of concentration. Vermeulen *et al.* (2008) also point at a high level of concentration in food retailing. Four retailers dominate the South African food retail sector, accounting for approximately 70% of the total grocery turnover. Supermarket outlets of three of these groups have been differentiated in order to cope with the different demands of different market segments; these different store types have in some cases been established in different city parts.

The stricter quality requirements of the retail trade, together with a desire to be able to deliver consistent services, have contributed to a move by both the retail groups and manufacturers to move away from product procurement from the municipal auction markets, which were the traditional spot markets in South Africa. The move has been toward vertical integration, either in the form of contracting arrangements or integration in ownership or management. Vermeulen *et al.* (2008) estimate that the four retail groups obtain the following percentages of their fresh produce from the traditional sources: 10%; 30%; zero; 20%.

Various authors have pointed out that integration, including contracting can be a major factor reducing transaction costs for producers and traders alike (Delgado, 1999; Ortmann, 2001; Sartorius and Kirsten, 2002).

It is however often very difficult for small producers, manufacturers, agents and traders to become part of these networks, for both local marketing and export. According to Kirsten and Sartorius (2002), it is often only the skilled, well endowed operators that are able to partake in these marketing chains and alliances; the standards set by the requirements of the modern consumers and retailers act as effective barriers to participation by small operators.

This does not, however, necessarily apply only to specialised or niche products, for example organic produce, hydroponic products, free range poultry, geographically differentiated produce, and the like. These may enable smallholders to move into the production of higher value goods. Smaller operators may well be able to serve such markets; specialised knowledge, capital items and skills are often prerequisites for such participation – traits which small operators with a history of poverty are normally not well endowed with. It may also require more capital than that is often available to small operators.

Smallholders tend to be excluded. In Africa, including large parts of South Africa, systems of communal land tenure very often make it impossible to use land as collateral for credit, thus adding to capital problems and hence, problems regarding market participation.

6.3 Contracting as a means to overcome barriers to market entry

Delgado (1999) mentions four keys to increased smallholder market access in Sub-Saharan Africa: access to assets, access to information, access to services, and access to remunerative markets. There are huge gaps in these fields all over Sub-Saharan Africa, including smallholder areas in South Africa. In addition, communal land tenure plays a role in this regard.

The general experience is that State-run efforts with regard to the above have not been successful or sufficient in Africa, including South Africa. It is obviously an area in which private initiative, where possible with government assistance, is needed. Success will furthermore be possible only if there is some degree of coordination in supplying assets, knowledge, services and market access. These have to aim at the same goal, or wastages will occur. There may otherwise also be wastages linked to opportunistic behaviour.

The only solution to the problem probably lies in contracting. Delgado (1999) advocates a system of interlinked input supply, credit and output marketing contracts and avers that these may be necessary to overcome moral hazard.

In Africa, successful smallholders have mostly been operators who have entered contracts with input supplies and/or for sale of outputs. Typical arrangements involve contract farming, producer co-operatives and outgrower schemes with or without a nucleus estate (Delgado, 1999; Glover, 1994; Grosh, 1994; Hussi *et al.,* 1993; Jaffee and Morton, 1995; Little and Watts, 1994; Swegle, 1994).

According to these authors (summarised by Delgado, 1999) these contracts typically involve agreements on selling the product of a specific area of a crop (often specified in terms of quality standards); this contract is usually made at planting time. The contract often specifies inputs and production practices, and inputs are usually supplied to the farmer; the costs of these inputs are redeemed at the time of crop delivery. The farmer has the benefits of an assured purchase and assured price as well as the necessary inputs; he typically also receives extension services from the contracting business entity. These extension services have in the overwhelming majority of cases been far superior to government supplied extension.

The processor/marketer naturally gains the advantage of an assured supply of the product with the desired traits at a previously arranged price. This facilitates further movement of the product down the supply chain. It has been mentioned that prices may at times be somewhat lower than prices at parallel markets (Badiane, 1997, quoted by Delgado, 1999).

There are however also certain possible problems with contract farming. In years with unfavourable production conditions (floods, drought, heat waves, cold spells, etc.) other market outlets may yield higher prices, and this may prompt producers to break their contract conditions and sell elsewhere. If the processor/marketer has supplied inputs to the producer, he may resort to legal action to retrieve the related costs. Such action may harm relationships with the community involved, not only with the culprit(s). The chances are that the processor/marketer will not be willing to enter contracts with the culprits in future; they may have forfeited future advantages for the sake of immediate gain. If the processor/marketer enters contracts not with individual farmers but rather with groups – e.g. local cooperatives – a free rider problem may manifest itself.

One may also find situations in which local relationships – e.g. local politics or ambitions – sow discord in a community and that such discord filters through to disharmony which eventually puts an end to a contracting scheme which has benefited at least some members of the community – if not the community as a whole.

Problems of moral hazard can thus not be ruled out. It is nevertheless essential that contract deals be based on mutual trust. Both sides to the agreement ought to do their utmost best to establish and strengthen such trust relationships.

The fixing and negotiation of prices may present problems. There are typically asymmetric power relationships involved. In cases where there is only one potential buyer – as is often the case – the buyer is in a monopsonistic position, and can recite conditions. This may eventually lead to situations of collapse of trust, end of the contracting arrangements and collapse of a scheme. The danger of such an occurrence depends very much on the processor/marketer's competitive position on his selling side and his access to alternative sources of products.

Contract farming has a long history, and has had a very chequered record in developing countries. Kirsten and Sartorius (2002) lists a large number of publications attesting to these two facts. Both successes and failures have abounded.

Another point of importance for South Africa is that although contracting with black farmers has been done in many sectors, the volume of supply from this source has been a small portion of the total contracting supply (Vermeulen *et al.*, 2008).

Based on the writings of Delgado (1999), Porter and Howard (1997) as well as Kirsten and Sartorius (2002), one may regard the following as guidelines for success for contracting to succeed with smallholder farming communities:

1. Educated people from the area who speak the predominant language(s) of the farmers, should be involved with senior management.
2. Interests of the farmers should be well represented in contract negotiations; farmer cooperatives may in this regard play an important role, provided free riding is not allowed.
3. Products requiring labour-intensive production are preferable. It is in these endeavours that smallholders are more likely to be competitive with larger units which may in the production of less labour-intensive products derive scale economies through more efficient mechanisation.
4. Crops that display a high value per hectare should be selected.
5. Payments should be directed to those who do the work. These are often women, and men sometimes object to such an arrangement. In this sense, Delgado refers to a case reported by Staal *et al.* (1997) in which such a matter was dealt with successfully in an ingenious, practical way.
6. Both the farmers and scheme management should be involved with monitoring the standards.
7. The wages and working conditions of hired workers should receive serious attention both in the setting up of the contracts and the monitoring thereof.
8. Participating farmers should be free to also grow other crops, particularly food, outside the scheme. This should relate to home consumption and also to sales.
9. Contractual relations should be based on mutual trust; contract manipulation by the agribusiness firm will almost inevitably lead to the collapse of the scheme.
10. The relationship should preferably span a wide spectrum relating to inputs, facilities and services. Improvement of the farmers' general business standards and knowledgeability should be a general goal in these schemes.

6.4 Experience with contracting: case studies

This section of the chapter will review some experiences with contract farming involving smallholders, as reported in South African publications and theses.

6.4.1 Smallholders' production of organic produce in the KwaZulu-Natal Embo Community

This part is selected from chapters in a book edited by Hendriks and Lyne (2009), in particular chapters contributed by Darroch andMashayanyama (2009); Hendriks *et al.* (2009); Lyne and Hendriks (2009) and Lyne *et al.* (2009a,b).

The organic food market is expanding rapidly, particularly in the developed world, and some South African food markets have also started marketing organic products. Organic production is also increasingly regarded as a plausible system for sustainable agriculture (Hellin and Higman, 2002) and it is also suitable for smallholders. Organic produce is however a niche product and unless special action is taken, smallholders are often excluded from the markets therefore. Success stories in some parts of Africa point at the need for collective marketing in accessing high-quality markets.

This approach was followed in the Embo community, situated approximately 40 kilometres south-west of the coastal city of Durban, South Africa. It is characterised by communal land tenure administered by traditional authorities. It is a rural area in a moist hinterland region with favourable climate for growing a wide array of crops throughout the year (Camp, 1995). In 2001, the Ezemvelo Farmers' Organisation (EFO) was established under the guidance of a professor at the University of KwaZulu-Natal with the object of improving the production, quality and marketing of the members' products. *Pro bono* organic certification was negotiated with AFRISCO, an accredited South African certifying body.

EFO pools *amadumbe (taro),* baby potatoes, and sweet potatoes grown individually by its members and sells them to a pack house which markets organic produce to a major retail chain. The contract is verbal. Thus, EFO is contracted to only one buyer. EFO started with 48 members and a contract to supply fresh produce to Pick'n Pay Stores Ltd. Members were given training in organic production. EFO is run by a Management Committee, which acts very much like a board of directors. By 2004, the membership had expanded to 151 members. The initial 48 members were fully certified organic producers and the other 103 partially certified, having fulfilled EFO's requirements and been presented to AFRISCO for full certification.

In an analysis of participation in EFO, Lyne *et al.* (2009b) statistically compared non-EFO members with partially and fully certified members. They found that neither gender, age nor literacy significantly determined membership. The coefficients obtained for size of farm were not statistically significant, but yet there was some evidence that households with more land tended to opt for less intensive, more conventional farming systems. An increase in benefits increased the likelihood of membership. Partially certified members perceived net benefits, despite the costs and meager returns experienced during the period of conversion – a similar result to that reported by Daminiani (2003) in Latin America and the Caribbean.

It also appeared that the likelihood of participating fully in EFO increased as the share of farming income in household income increased. EFO members who rated grazing livestock as a serious threat to their crops were more likely to be partially than fully certified; this may relate thereto that fencing materials had been donated to the early adopters, but not to those who joined later.

It was also found that the odds of fully certified membership increase when the participant regards a lack of affordable operating inputs as a serious problem.

The potential problem with free-riding also received attention. Nabli and Nugent (1989) was quoted that asymmetric information and a lack of mutual trust contribute to opportunistic behaviour, raising transaction costs and encouraging free riding In the present case, the agreed upon grading procedures agreed upon between EFO and the pack house has been flawed as the produce cannot be traced to its point of origin as it is pooled before being graded by the pack house. Individuals who are aware thereof may thus deliberately channel inferior products through the pack house, and analysis showed that this had actually happened. Certification status was found to significantly influence free riding; new (partially certified) members were more likely to free ride than founding, fully certified members. This is consistent with findings elsewhere that length of association is negatively associated with free-riding (Chong, 2001).

The EFO experience can in general be said to have been successful and to have raised welfare in the community. There is, however, room for improvement. One relates to traceability of products to the origin, rather than pooling before grading. It also appears that EFO should target households that rely on farming for income and that are relatively land-constrained. One can expect more enthusiastic cooperation from such members. Some members are not completely satisfied with the pricing mechanism, and feel that with a single buyer, they get disadvantaged from asymmetric power relationships – the age-old problem of monopsonistic behaviour.

Ranking was done of the top 20 constraints cited by members on the competetiveness of the Embo supply chain (Darroch and Mushayanyama,, 2009). The top six constraints, in order of importance, were: uncertain climate; unavailability of tractors when needed; delays in paying for products sent to pack house; inputs not available at affordable prices; lack of cash and credit to finance inputs; and lack of affordable transport. Nothing can be done concerning climate; the other matters should receive attention.

6.4.2 The Swaziland sugar industry

This part is derived from Masuku and Kirsten (2004); Masuku *et al.* (2003, 2007). The Swaziland sugar industry derives its present structure from the Sugar Act of 1967. Under this legislation, sugar millers as well as growers need licenses or quotas to mill sugar or produce sugarcane. Cane growers are responsible for growing cane and they bear the costs of cane delivery to the mills. They are subject to delivery quotas in terms of the weight of sucrose in the cane delivered. They are required to deliver their full quota to a designated mill, and the mill is required to accept all cane delivered to it up to each grower's quota. If a grower delivers more cane, a lower price is paid for the excess amount.

The system creates the potential for opportunistic behaviour, as each miller is in a monopsonistic situation. Opportunistic behaviour could be aimed at exploitation of 'quasi-rents' generated by committed capital which once it has been invested, cannot readily be used for another purpose. In the Swaziland sugar industry, the buyer has monopsony power and he can bargain the price of cane down. As he has already committed his capital, the grower is still better off continuing production and delivery than by closing down.

This does of course, not relate to prospective new growers. In the long run, the survival of the system will depend on the level of cooperation – and trust – between cane growers and millers.

Statistical modeling has in the case of growers and millers of sugarcane in Swaziland shown the following seven hypotheses to be valid:
1. Growers' trust in the relationship with millers is negatively influenced by their perception of opportunistic behaviour on the part of the miller.
2. Growers' trust level in their relationship with millers is positively associated with their level of cooperation.
3. Growers' level of trust is positively associated with their level of satisfaction with the relationship.
4. Higher trust levels lead to higher levels of cooperation and hence also to higher levels of commitment in the relationship by growers.
5. Cooperation between growers and millers will positively influence benefits accruing to growers.
6. Farmers' realisation of the benefits from the business relations enhances their satisfaction with such relationships.
7. Farmers' perceptions about the millers' level of cooperation directly affect their levels of satisfaction.

This clearly illustrates the need on the part of buyers (millers in this case) both to be honest in their dealing and to maintain sound public relations with their suppliers. This remark is borne out by one other analysis which indicates that the cane growers are certain about and committed to the relationship by millers, but they do perceive poor cooperation between the two sides. There is also a perception of opportunistic behaviour on the millers' part. So, although they still trust the millers, it is to a limited extent; they feel that they are dependent on the millers as they are locked into the mill.

6.4.3 Malt barley at Taung irrigation scheme

This part was largely taken from the masters' thesis by Klopper (2009) at the University of the Free State. All references, except Tregurtha and Vink (1999) appear in the thesis.

South African Breweries (SAB), now known as SABMiller (SABM), is by far the largest brewery firm in South Africa, and is also a major player on the international beer and brewery industry. SAB and SABM have for long had to import most of the barley used from malting, since most locally grown barley – mostly on dry land in the Southern Cape (part of Western Cape Province) – has a relatively poor malting quality. Largely because of the depreciation of the Rand on currency markets and the general volatility of currency markets, SAB invested in research and development (R&D) to develop and establish production of better malting quality barley cultivars in South Africa. By 1994 the R&D programme had made sufficient progress, and SABM started an effort to establish production of the selected cultivars under irrigation. The Taung Irrigation Scheme was selected for this purpose (Tregurtha and Vink, 1999).

The Taung Scheme is situated in North West Province and was started during the 1930's on tribal land when almost 2,000 farmers, selected by tribal chiefs, were settled on 1.7 ha plots. Subsistence farming was the order of the day, and maize and pumpkins were the preferred crops. The unavailability of a ready supply of inputs, a lack of credit, water logging and salinity caused many plots to be abandoned and overtaken by weeds. Later efforts, starting in 1974, to redevelop the scheme, settling farmers on larger plots (20 ha each) have met with rather limited success, largely because of organisational weaknesses, including lack of credit facilities (farmers cannot offer land as collateral), and lack of marketing infrastructure, equipment, appropriate extension and a well organised local economic structure. A serious drought in 1984 which caused water allocations to be cut by 50%, added further hardship. The scheme did not succeed to bring welfare to the participants. Such was the situation when SABM, together with the North West Department of Agriculture, established a barley growing project in Taung.

SABM performs a dual role: it acts both as a financial intermediary and as contract buyer of barley. Compared to most commercial banks, agri-industrial firms often have an advantage in supplying smallholders with credit. The close relationship SABM has with its contract farmers lowers the risk of default, allowing the firm to lend money to growers unable to obtain loans from commercial banks. The risk is lower for SABM for the following reasons:

1. credit is supplied in the form of inputs;
2. SABM monitors the use of the inputs, so the inputs cannot be resold;
3. entry barriers to markets limit growers' ability to sell the product in markets other than SABM;
4. unlike a commercial bank, SABM is able to acquire repayment directly from the crop revenue; and
5. the farmer knows that default on his part will prevent him from acquiring loans in future.

SABM, particularly its Agricultural Services Department, plays an ongoing role in supplying a host of services. The most important are probably the following:

- information and report back meetings days ahead of the planting season;
- negotiations with suppliers of inputs (fertiliser, pivot maintenance and chemicals) and contractors (soil preparation, planting, fertiliser applications, weed control, etc.) in order to bargain for the best prices for producers;
- written contracts with individual producers;
- contracts for each individual farmer with Electricity Supply Commission (Eskom);
- soil sampling;
- negotiations to update farmers' insurance policies and obtain the best tender;
- coordination of production activities; and
- marketing of feed grain barley.

The farmers are allowed to plant alternative crops, and other crops are grown. Lucerne (alfalfa), is mostly sold on the farm itself; farmers have problems with rising prices for inputs for lucerne production (Gronum *et al.*, 2000). There are presently efforts by the Vaalharts Cotton Growing Association to establish small commercial cotton growers at Taung by establishing contract ginning services or forward integration in farmers' own ginning. Some wheat is also grown for home consumption.

The farmers have to perform the barley production activities, e.g irrigation, assistance with soil sampling, regular inspection of fields, etc.

The project began with 55 farmers and by the end of 2004, the number of farmers had grown to 178 with 3,500 ha of land under barley, using sprinkler and pivot point irrigation systems. The project can easily be expanded to cover 5,200 ha (BFAB, 2008). By the end of the 2004 production season, the Taung Scheme supplied 10,000 tons of barley (almost 24% of the 42,000 tons SABM needs for its South African operations), earning in excess of R20 million – a large increase from the R2.4 million in 1998 (BFAB, 2008). This indicates a large degree of success in a scheme which had not previously had a very happy history.

Statistical analyses were performed to analyse the main factors leading to potential success at Taung. Having a contract with SABM, access to electricity and to credit have clearly been shown to be the biggest factors positively influencing potential success for the smallholder irrigators. A negative relationship between success potential and market distance was found for farmers who have not been contracted to SABM, but not for those with SABM contracts; for these farmers, the influence of distance is minimal as the barley is collected by contractors from SABM.

This case study demonstrates a positive outcome of contracting. Care will however have to be taken to prevent the asymmetric power relationship from affecting the relationship negatively at a later stage. In the 1990's, relationships between SAB and contracted commercial hops growers in the George area of the Southern Cape was strained for a while. The possibility of cotton delivery and management contracts may in future also reduce the asymmetric nature of power relationships.

6.5 Recommendations

Some conditions have be fulfilled if contracting is to contribute meaningfully to the access of small farmers to profitable and lucrative markets. It will be necessary to overcome the existing or potential pitfalls in this kind of arrangement, and to create an environment in which this type of arrangement will have a high probability of success.

6.5.1 Counteracting monopsonistic situations

Firms entering contracts with small-scale producers can potentially assume monopsonistic positions and should these firms choose to exercise their monopsony powers, the smallholder farmers will inevitably land in a position unfavourable to their welfare and development.

Countervailing power is potentially the most effective way to counteract this problem: the producers (i.e. smallholders) can form a bargaining body which can negotiate contracting arrangements and conditions with the firm which buys the products. This bargaining body must be independent of outside regulation, particularly by tribal authorities. It should be able to represent all interested producers of a particular product in the specified area; such a body can negotiate for exclusive delivery in the area, thereby preventing free riding.

Local cooperatives may be the appropriate type of institution to handle this task. Care must be taken that such cooperatives truly represent their members and that factional interests do not preponderate in the cooperative. Citing other literature in this regard, Fekete *et al.* (1992) state that if some leading members are more powerful than lower level members, they may use the cooperative for personal gain to the detriment of the others.

If a cooperative is formed for this purpose, it will also be necessary to avert the pitfalls which have caused the collapse of many cooperatives in developing areas, as described by Machethe (1990). The cooperative must be free from any governmental interference; its members must understand its goals and actions and must be involved in its decision-making; and it must be well managed.

Governmental agencies ought to be established to foster cooperatives for smallholders; the role must be that of advice to management and members alike without any direct interference in the business actions of the cooperative. The cooperation of private sector firms and NGOs in this type of venture will be needed for success.

6.5.2 Niche products

Experience has often shown that farmers are often more inclined to adopt new practices in the production and marketing of products new to them than in those which they have been producing for a long time. In the modern marketing world with increased emphasis on quality products, this opens the opportunity for niche product production and contract marketing of those products, as illustrated by the experience of the Embo Community in KwaZulu-Natal. In this particular case, a community which has had rather limited experience in commercial farming has been activated into what has thus far been a successful conversion, leading to the potential of better living standards.

A prerequisite for success will probably be a buying concern with a standing in the marketing of the particular product or group of products.

Possibilities for converting some 'regular' products into niche products may also not be as difficult as one may think, provided some group or firm takes an initiative and succeed in organising a community. The possibilities are certainly wide. One may for example think of a niche cheese made from from a particular breed of cattle (or goats, or sheep) from a specific community in a specific area; Flowers can also provide a good field fo such a move.

6.5.3 Individual accountability

Farmers differ in the extent to which they will always be able to deliver products of quality, and if a system does not exist to trace quality back to the individual producer, those with less success in delivering high quality products may reduce the results for the deliverers of consistently high quality produce. This matter should receive serious attention in contracting schemes. This will probably necessitate quality assessment at thee time and place of either delivery by the producer, or at the point of collection.

6.5.4 Credit and services

Small-scale farmers, particularly those in a communal ownership situated have experienced serious problems with the availability of production credit. In certain cases, such as Taung, this has been supplied by the contracting company. This particular case study pointed at some important advantages of this approach. However, not all contracting firms have the organisation or desire to be involved in the credit side of the business. Contracting farmers' organisations, e.g. cooperatives formed by them, may however be in a position to arrange with financial institutions to supply the needed credit on perhaps a group credit system, as discussed in Chapter 8.

A similar problem exists with respect to services such as ploughing, pest control, transport, harvesting, etc. This work can often be done by means of contracting. Those involved with training, information oand mentoring of these farmers should play a supervisory role in this regard.

6.5.5 Information and mentorship

Commercial production is in itself a new venture for many small farmers, who do not have much experience of the commercial farming world. Many have also had rather limited formal education, and new products as well as new technology present huge challenges. The farmers have also not been schooled in business affairs, nor have they had experience in modern business practices. This all points to a need for information – such as a high quality agricultural extension – which in the light of the existing weakness of state-run extension services, will need inputs from the private sector. Experiences mentioned in this chapter point at the success which can be obtained when the buying firm renders this service. However, how certain can one be of the permanency of such an arrangement? There is certainly scope for arrangements in which the farmers' associations cooperate with the private sector to ensure ongoing information transfer to these new commercial producers.

Organised agricultural bodies have in some agricultural industries achieved considerable success with mentorship programmes serving emerging commercial farmers. There is room for expansion of these services also for new entrants to the contracting farming scene.

6.6 Conclusion

Contract farming has, as stated before, had a rather chequered history over the world, including South Africa. Failures appear to be mostly been caused by poor organisation, opportunistic behaviour and lack of trust. These factors can be restored by good faith, honesty and good organisation. Given the dire necessity for development of smallholder agriculture, particularly in the developing world, efforts ought to be done to foster contract smallholder agriculture in an era when large numbers of consumers become more discerning, thereby also creating more opportunities for special products.

References

Badiane, O., 1997. Libéralisation et compétitivité de la filière arachidière au Sénégal. Market and Structural Studies Division Discussion Paper 17. International Food Policy Research Institute, Washington, DC, USA.

Bureau for Food and Agricultural Policy (BFAB), 2008. Small-scale farmer's brief: macroeconomic impacts on profitability. BFAB Brief September 2008. Available at: http://www.bfab.co.zs.

Camp, K.G.T., 1995. The bioresource units of KwaZulu-Natal. Report No. N/A/95/32, KwaZulu-Natal, Department of Agriculture, Cedara, South Africa.

Chong, A., 2001. What are the determinants of free riding? Inter American Development Bank, Washington, DC, USA.

Daminiani, O., 2003. The adoption of organic agriculture among small farmers in Latin America and the Caribbean: thematic evaluation. International Fund for Agricultural Development, Rome, Italy.

Darroch, M.A.G. and T. Mushayanyanyama, 2009. How can the competetiveness of the Ezemvelo farmers' organisation supply chain be improved? In: S.L. Hendriks and M.C. Lyne (eds.) Does food security improve when smallholders access a niche market? Lessons from the Embo Community in South Africa. The African Centre for Food Security, Pietermaritzburg, South Africa, pp. 93-105.

Delgado, C.L., 1999. Sources of growth in smallholder agriculture in Sub-Saharan Africa: the role of vertical integration of smallholders with processors and marketers of high value-added items. Agrekon 38: 165-189.

Delgado, C.L. and A. Siamwalla, 1997. Rural economy and farm income diversification in developing countries. Paper presented at a plenary session of the XXIII International Conference of Agricultural Economists, Sacramento, CA, USA.

Drucker, P.F., 1958. Marketing and economic development. The Journal of Marketing 22(3): 252-259.

Fedderke, J. and G. Szalontai, 2005. Industry concentration in South African manufacturing: trends and consequences, 1972-1996. Working paper series No.96, School of Economics, University of Cape Town, Cape Town, South Africa.

Fekete, F., T.I. Fenyes and J.A. Groenewald, 1992. Some forces affecting performance and organization of agricultural cooperatives. South African Journal of Economic and Management Sciences 8: 16-23.

Glover, D., 1994. Contract farming and commercialization of agriculture in developing countries. In: J. Von Braun, and E. Kennedy, E. (eds.) Agricultural commercialization, economic development and nutrition. The Johns Hopkins University Press, Baltimore, MD, USA.

Gronum, C.F., H.D. Van Schalkwyk, A. Jooste, J.H. Du Plessis, F.I. Geldenhuys, D.B. Loouw and J.P. Geldenhuys, 2000. Lucerne production in South Africa. Research report. University of the Free State, Bloemfontein, South Africa.

Grosh, B., 1994. Contract farming in Africa: an application of the new institutional economics. Journal of African Economics 3: 231-261.

Hellin, J. and S. Higman, 2002. Smallholders and niche markets: lessons from the Andes. ODI Agricultural Research and Extension Network Paper 118, ITDG Publishing, Oxford, UK.

Hendriks, S.L. and M.C. Lyne (eds.), 2009. Does food security improve when smallholders access a niche market? Lessons from the Embo community in South Africa. The African Centre for Food Security, Pietermaritzburg, South Africa.

Hendriks, S.H., Lyne, M.C., Katunda, M. and Mjonono, M., 2009. Description of the Ezemvelo Farmers' Organisation, the survey methodology and household demographics. In: S.L. Hendriks, and M.C. Lyne, (eds.) Does food security improve when smallholders access a niche market? Lessons from the Embo Community in South Africa. The African centre for Food Security, Pietermaritzburg, South Africa, pp. 19-28.

Hussi, P., J. Murphy, O. Lindberg, and L. Brenneman, 1993. The development of cooperatives and other rural organizations. The role of the World, Technical Paper No. 199, African technical Department Series. World Bank, Washington, DC, USA.

Jaffee, J. and J. Morton, 1995. Marketing Africa's high-value foods: comparative experience of an emerging private sector. World Bank, Washington, DC, USA.

Kirsten, J. and K. Sartorius, 2002. Linking agribusiness and small-scale farmers in developing countries: is there a need for contract farming? Development Southern Africa 19: 503-529.

Klopper, J.P., 2009. Mainstreaming of smallholder irrigators: the case of Taung Irrigation Scheme, North West, South Africa. MSc (Agric.) thesis, University of the Free State, Bloemfontein, South Africa.

Little, P.D. and M.J. Watts, 1994. Living under contract: contract farming and agrarian transformation in Sub-Saharan Africa. The University of Wisconsin Press, Madison, WI, USA.

Lyne. M.C. and S.L. Hendriks, 2009. Did food security improve when smallholders assessed a niche market for organic produce? In: S.L. Hendriks, and M.C. Lyne, (eds.) Does food security improve when smallholders access a niche market? Lessons from the Embo Community in South Africa. The African centre for Food Security, Pietermaritzburg, South Africa, pp. 133-138.

Lyne, M.C., S.L. Hendriks and J.M. Chitja, 2009a. Agricultural growth and food security. In: S.L. Hendriks, and M.C. Lyne, (eds.) Does food security improve when smallholders access a niche market? Lessons from the Embo Community in South Africa. The African centre for Food Security, Pietermaritzburg, South Africa, pp. 1-10.

Lyne, M.C., S.L. Hendriks and L. Gadzikwa, 2009b. Is the Ezemvelo Farmers' Association serving its members? Organisational and contractual challenges. In: S.L. Hendriks, and M.C. Lyne, (eds.) Does food security improve when smallholders access a niche market? Lessons from the Embo Community in South Africa. The African centre for Food Security, Pietermaritzburg, South Africa, pp. 72-92.

Machethe, C.L., 1990. Factors contributing to poor performance of agricultural co-operatives in less developed areas. Agrekon 29: 305-309.

Makhura, M.T., 2001. Overcoming transaction costs barriers to market participation of smallholder farmers in the Northern Province of South Africa. PhD thesis, University of Pretoria, Pretoria, South Africa.

Masuku, M.B. and J.F. Kirsten, 2004. The role of trust in the performance of supply chains: a dyad analysis of smallholder farmers and processing firms in the sugar industry in Swaziland. Agrekon 43: 147-161.

Masuku, M.B., J.F. Kirsten and R. Owen, 2007. A conceptual analysis of relational contracts in agribusiness supply chains: the case of the sugar industry in Swaziland. Agrekon 46: 94-115.

Masuku, M.B., J.F. Kirsten, C.J. Van Rooyen and S. Perret, 2003. Contractual relationships between smallholder sugarcane growers and millers in the sugar industry supply chain in Swaziland. Agrekon 42: 189-198.

Nabli. M.K. and J.B. Nugent, 1989. The new institutional economics and development: theory and application to Tunisia. Elsevier Science, New York, NY, USA.

Ortmann, G.F., 2001. Industrialisation of agriculture and the role of supply chains in promoting competitiveness. Agrekon 40: 459-489.

Porter, G. and K.P. Howard, 1997. Comparing contracts: an evaluation of contract farming schemes in Africa. World Development 25: 227-238.

Sartorius, K. and J. Kirsten, 2002. Can small-scale farmers be linked to agribusiness? The timber experience. Agrekon 41: 295-325.

Staal, S, C. Delgado and C. Nicholson, 1997. Smallholder dairying under transaction costs in East Africa. World Development 25: 779-794.

Swegle, W.E. (ed.), 1994. Developing African agriculture: new initiatives for institutional coordination. Proceedings of a Workshop in Cotonou, Benin. Sasakawa Foundation, Tokyo, Japan.

Tregurtha, N.L. and N. Vink, 1999. Trust and supply chain relationships: a South African case study. Agrekon 38: 755-765.

Vermeulen,, H., J. Kirsten and K. Sartorius, 2008. Contracting arrangements in agribusiness procurement practices in South Africa. Agrekon 47: 198-221.

Vojtech, V., H. Huang, O. Melyukhina and P. Liapis, 2006. OECD review of agricultural policies: South Africa. OECD Publications, Paris, France.

7. Food retailing and agricultural development

Lindie Stroebel and Herman D. van Schalkwyk

7.1 Introduction

The global food retail sector has, over the past few decades, evolved into a highly dynamic and competitive industry. The rise of supermarkets, as explained by many other researchers, was initially experienced in developed countries in Northern America and Western Europe (Cacho, 2003; Hagen, 2002; Reardon and Berdegue, 2002; Senauer and Venturini, 2005; Weatherspoon and Reardon, 2003). The majority of supermarkets spread to less-developed and developing countries, as economic, political and social environments allowed, and multinational companies wished to expand beyond their saturated and highly competitive domestic markets. In an overview of the rise and development of the supermarket industry, as diffusion model, Reardon *et al.* (2003) observe a shift in the tendency of supermarkets to occupy only a small niche in capital cities, serving only the rich and middle class, to spreading well beyond the middle class in order to penetrate deeply into the food markets of the poor.

Over the last 35 years, developed countries experienced rapid growth in both size and numbers of supermarket chain stores. Reardon and Berdegue (2002) state that supermarkets are now the dominant players in the agrifood economies of many developing countries, specifically emphasising the fact that supermarkets have become the dominant food retailers in some African countries over the past decade. The influence exerted on local economies, the traditional retail sector and the agricultural sectors are diverse and significant.

The primary objective of this chapter is to inquire into the evolving supermarket industry in South Africa, to consider the effect on local economies, traditional food retailers and the agricultural sector. In an attempt to achieve the primary objective, it is inevitable to depict the determinants of expansion and development of supermarkets in the world, to understand the observed expansion trend of the international food retail industry. Accordingly, the impact of supermarket expansion on regional economies, the traditional and/or alternative market place and the agricultural sectors in Africa, can be debated with reference to previous findings.

7.2 Supermarket diffusion into Africa

The development of the international food retail industry arose from expansion by the supermarket industry in countries such as Western Europe and the United States of America. Spill-over started occurring to directly adjacent countries. The more drastic development, due to expanded spill-over from developed countries, can be explained in four phases. The first supermarket entrants and introduction initiatives in the developing world took place in the larger and richer Latin American countries. Trail (2006) specifically found in his projection for supermarket penetration that projected income growth will have a significant impact on supermarket penetration in middle-income Latin America and transition countries. The second phase happened in East and South-East Asia, where Trail (2006) actually found that projected income increases have little impact in the poorest countries like Bangladesh and Pakistan. Urbanisation and incomes again have a significant impact on supermarket penetration in China (Trail, 2006). This is followed by the third phase, in Southern and East Africa, as well as the smaller and poorer Latin American countries. The final phase is still in process, as development in the food retailing environment is taking place in secondary cities and towns in Latin America and the South-East and East Asian countries. Also included in this phase is the newly targeted South Asian and West African countries, as well as the secondary towns and rural areas in Southern and East Africa. The phases of this diffusion model are graphically in Figure 7.1.

Africa is a much smaller role player in international food retailing, but the rise of supermarkets is directly related to westernisation and urbanisation trends currently

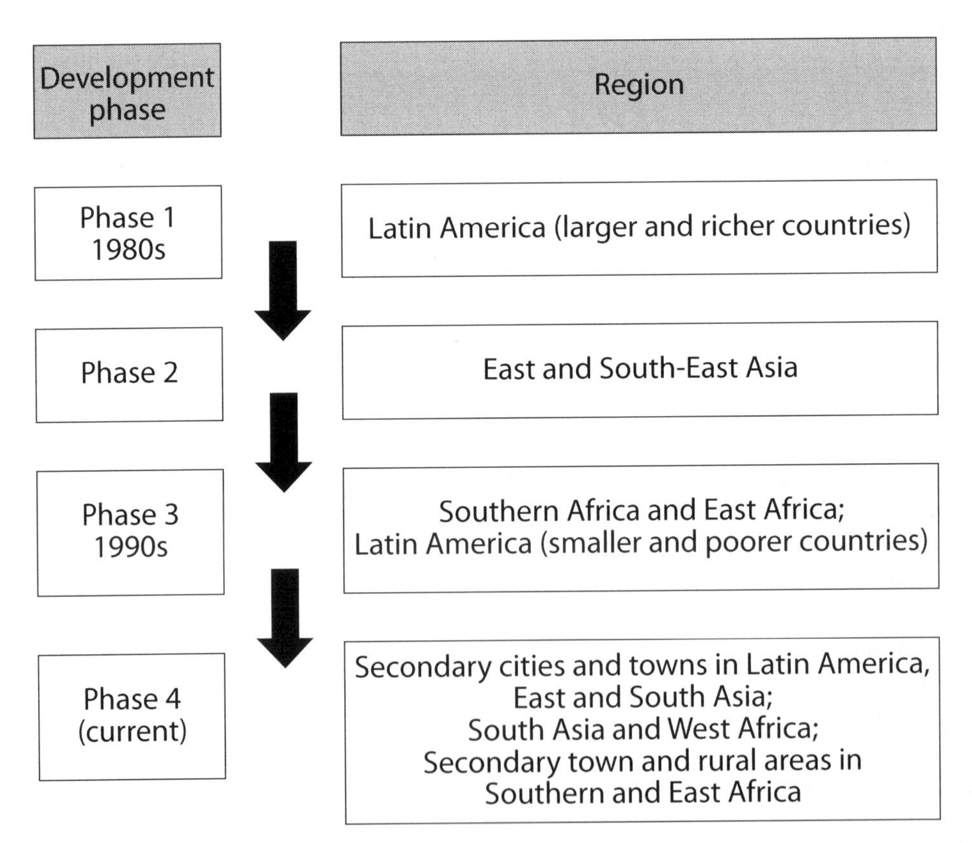

Figure 7.1. Phases of the diffusion model of the supermarket industry in the developing regions of the world (based on Reardon et al.*, 2003; Humphrey, 2007; Goldman* et al.*, 2002).*

experienced throughout Africa. Trail (2006) found that the continued liberalisation of inward foreign direct investment, resulting as it does in competition with and/or entry of multinational retailers, is likely to be the main driving force for the continued spread of supermarkets in developing countries. The forces of globalisation and urbanisation across the developing world are an inevitable reality. An ever-increasing number of city-dwellers depend on supermarkets as their main food source. Changing consumer behaviour could evolve in a trend that will benefit the supermarket and other dynamic food retail markets in becoming significant international role players.

The development of the supermarket industry in Africa is mostly characterised through the rapid rise of supermarkets in South Africa and Kenya and their spill-over into other countries in recent years. The initial signs of supermarket development occurred in the larger and richer countries, where domestic chains grew in size and number. According to Weatherspoon and Reardon (2003), it then only expanded to smaller and poorer countries through FDI. The economic geography of expansion and spread of supermarkets was related to the nature of sub-regions/countries' economies, policies, political stability and purchasing power. The speed and cumulative acceleration was, according to Weatherspoon and Reardon (2003), much less predictable and extremely surprising. The supermarkets were initially focussed, through location and sales, on upper-income consumers in each respective country. In the 1990s the number of supermarkets increased rapidly, and diversified to various different outlet formats, such as hypermarkets and convenience stores. These retail formats replaced the more traditional retailers such as small shops and, in some cases, public markets. At this time the more traditional food retail outlets existed mainly in the rural areas.

According to Reardon *et al.* (2003) the frontrunner in the African context was South Africa, where a spectacular rise occurred after the end of apartheid in 1994. The South African supermarket industry exploded over the past decade. Many FDI opportunities became available to and from South Africa. Investment opportunities in the rest of Africa were boundless. Reardon *et al.* (2003) indicated that South African supermarket chains had already invested in 13 African countries. By the late 1990s, the supermarkets expanded to poorer and rural areas by means of franchises. According to *Business Today*, as cited by Weatherspoon and Reardon (2003), supermarkets were not permitted to exist in townships and former homeland areas during apartheid. Between 1994 and the early 2000s supermarkets focussed on consolidating business in the cities. Competitive pressure at the top end of the market pushed supermarkets to expand to townships, starting during mid-2001. Pick 'n Pay and Shoprite exerted a significant strategic push when they entered the rest of the African market through FDI, partly due to near saturation of South African markets, and partly because of the search for higher profits in other markets with fewer stores. This FDI incentive ignited a wildfire of investment by domestic chains in order to defend their market shares.

Kenya was the other front runner in the rise of supermarkets, followed by Zimbabwe and Zambia. In Kenya the majority of supermarkets are established in Nairobi, but due to further expansion, about one quarter of supermarket equivalents are now outside Nairobi. Supermarkets are currently being introduced in the medium-sized cities and larger towns.

Tanzania underwent a series of macroeconomic reforms and liberalisation of FDI in the 1990s and South African companies used the opportunity to enter that market. Two Kenyan chain companies also expressed their intentions to enter the Tanzanian market. The apparent opportunity for growth in Tanzania has led many supermarket chains to view the country as an important emerging market (*Business Day* 25 November 2002; Shoprite, 2004; Wahome, 2001; Wandera, 2002; Winter-Nelson, 2002; all cited by Weatherspoon and Reardon, 2003).

In summary, Table 7.1 contains a ranking of African countries regarding supermarket development (Weatherspoon and Reardon, 2003).

Table 7.1. Present stages of supermarket development in various countries in Africa.

Stage of development	Country
Recent entry	Ghana, Democratic Republic of the Congo
Small niche market	Tanzania, Nigeria, Uganda
Early-intermediate stage	Zambia, Zimbabwe
Intermediate stage	Kenya
Advanced stage	South Africa

7.3 Food retailing in South Africa

The history and development of the food retail industry in South Africa can be explained by describing the rise of the supermarket industry. Supermarkets started in South Africa in 1951 when OK Bazaars opened a food department store as its flagship store in Eloff Street, Johannesburg. Since then, many entrants have followed into the industry.

According to Weatherspoon and Reardon (2003) there are neither official data nor private estimates (such as those of AC Nielson) on how quickly the supermarket sector in South Africa has grown over the past decade. Generally it is perceived that some supermarket chains started quite early, but have grown rapidly only recently (in particular, since 1994). Pick 'n Pay started in 1967 and Shoprite-Checkers in 1979. Shoprite's major growth came with acquisition of some of South Africa's existing domestic chains, like the 17-store chain of Grand Supermarkets in 1990 and the 170-store chain of Checkers Supermarkets in 1991. SPAR originated in the Netherlands in 1932, but only came to South Africa by 1963.

Woolworths Food opened as part of a department store and has seen tremendous growth since its inception. All the major role players in the supermarket industry only experienced rapid growth over the last 10 years.

The supermarket was historically the first format used, with location and sales focused on upper-income consumers. While the number of supermarkets increased quickly during the 1990s, by the late 1990s the chains added hypermarkets (to extend their target to middle- and lower middle-class urban consumers, with broad food and non-food selections and low prices). In the late 1990s and early 2000s chains added convenience stores on transport routes and in dense urban areas. During the second half of the 1990s small supermarkets opened in poorer areas by means of franchising (Weatherspoon and Reardon, 2003). However, until the early 2000s, supermarkets were still focused on consolidating business in the cities. Competitive pressure and relatively saturated markets were experienced at the top end. This pushed some supermarket chains to expand into townships from 2001. Supermarkets were not permitted to exist in townships and former homeland areas during apartheid (*Business Today*, as cited by Weatherspoon and Reardon, 2003). The companies entered these regions by domestic capital acquisitions of smaller chains or independent stores located in those areas.

7.3.1 Formal food retailing sector

The formal sector represents a wide spectrum of neighbourhood convenience stores (include forecourts, described as gas stations with convenience type stores, as well as cafes, in the format of internationally known neighbourhood convenience drugstores), speciality stores, boutiques, chain supermarket stores, department stores and large wholesale and retail outlets. The various types of food retail outlets differ in size and in turnover. For food retail outlets to be considered for inclusion in AC Nielson data for the formal food retail sector, it:

- must stock at least four food and four confectionary product classes;
- must have a fixed location;
- must be plainly recognised as a trader; and
- must be accessible from any public thoroughfare.

Food retail outlets excluded from the total are school tuck shops, canteens, kiosks, hawkers and mobile stores, which are very difficult to count, due to various reasons. Makro, which is a wholesale food retailer, as a subsidiary of MassMart retailers, is also excluded due to the fact that Makro only sells canned and non-perishable foods and no fresh or perishable processed foods.

7.3.2 Informal food retailing sector

The informal food retail sector is playing a much larger role in the food retail industry than has been recognised before. Its size, volumes and revenues cannot be determined easily, as the outlets are not registered as retailers, nor do they pay rent or taxes. RocSearch (2004) estimated that the informal retail market had a turnover of roughly R34 billion in 2004. Wilson (2003) estimates that there could be over 6,000 spaza shops in South Africa.

Informal food retail stores are independent and include general dealers, small cafes, street vendors, hawkers, tuck shops, primitive street corner stalls and spaza shops. Most procurement by these dealers is from large cash & carry stores and wholesalers, while some products are bought from supermarkets and, to a smaller extent, from local producers. Spaza shops are the most common type of informal outlet and are usually found in townships and poor neighbourhoods.

A spaza shop is usually attached to the owner's house, but can be in a garage, an outside room, a shipping container or it could be a street vendor. Spaza shops, like many other informal retailers, find their success by offering the clients place convenience. Khuzwayo (2000) states that spaza shops have an advantage due to the fact that they are close to their customers and that their overheads are lower than conventional stores. Also in the favour of the spaza shop owner is the fact that he/she is often a neighbour of the clients.

Another familiar type of food retail outlet found in townships and other urbanised, low income areas are small counter-service stores, where a shop assistant stands behind the counter and passes the client whatever he/she asks for.

7.3.3 Dispersal of retail outlets in South Africa in 2004

Different retail outlets occur more abundant in different regions of the country (Figure 7.2). Throughout the country, the most common store types by number are the smaller types, such as rural shops, urban counter shops and urban self-services shops. Urban counter shops are usually situated in townships and other poor neighbourhoods, which explain the high frequency of occurrence. These smaller shops totally outnumber the large supermarkets and more formal retail outlets, described as total majors, branded superettes and forecourts. There are a large number of rural shops in all provinces, except Gauteng and Western Cape, which, together with KwaZulu-Natal, has large numbers of urban counter shops. According to the numbers of outlets, it appears that there are very few 'majors', such as supermarkets and hypermarkets, compared to other retail markets. Figure 7.2 indicates the dispersal of various outlets throughout South Africa, as was the case in 2004. It is expected to have stayed relatively the same over the past 4 years, but due to the unavailability of updated information this graph will give sufficient background as to the status of the number of different retail stores per region.

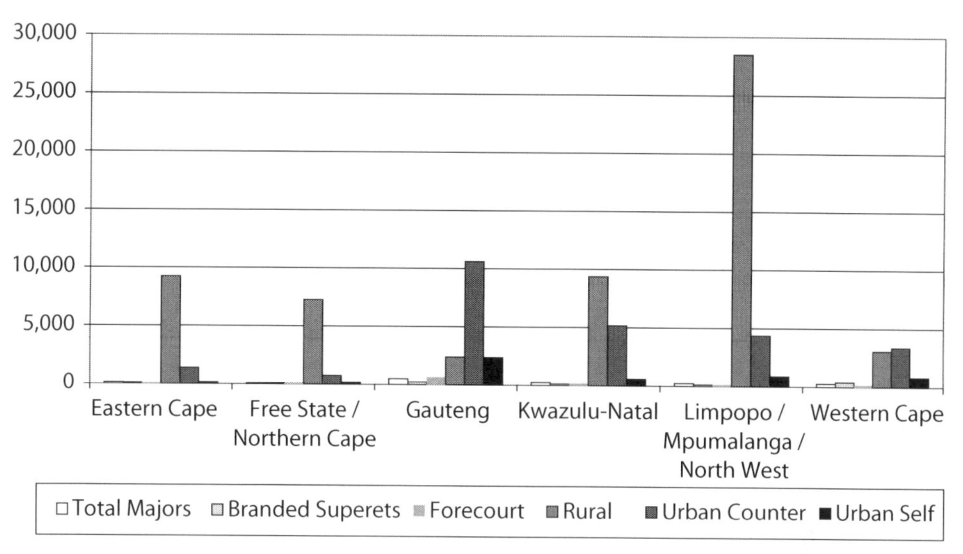

Figure 7.2. Numbers of different food retail store types per region in South Africa in 2004 (AC Nielson, 2004).

7.3.4 Size of the market

The South African food retail industry had a compound annual growth rate in the period 2003 to 2007 of 2.1%. Between 2006 and 2007 the industry grew slightly slower, with a growth rate of 1.5%, to reach a value of R153 billion (Datamonitor, 2008a,b).

Since the beginning of the this decade, rivalry among food retailers in South Africa became more and more fierce, especially with the rapid penetration of Supermarkets and Hypermarkets. According to data obtained from Datamonitor (2008a,b), supermarkets account for the largest share of the South African food retail industry, generating 55.6% of the industry's value, whereas food specialists account for 22%, hypermarkets for 8.3% and discounters for a mere 2%. Figure 7.3 indicates the industry segmentation as percentage share by value in 2007.

7.3.5 The South African supermarket industry

Pick 'n Pay supermarket group originated in 1967 when the first stores opened in Cape Town. Currently Pick 'n Pay is regarded as one of Africa's largest and most consistent retailers of food, general merchandise and clothing. Tremendous growth in turnover was experienced between 2001 and 2003, followed by a lower, though more stable rate of growth thereafter. Since 2004 a steady growth rate of slightly higher than 10% were maintained, with the exception of 9% in 2005. After a slight slowdown in annual growth in 2008, the growth rate hiked again in 2009 by 17.39%.The total annual turnover increased steadily to R49.8 billion by the end of the 2008/2009 financial year in February 2009. The Pick 'n Pay

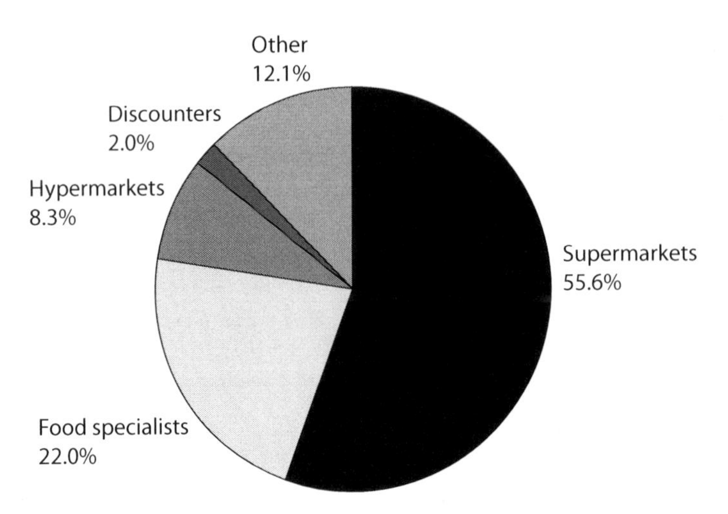

Figure 7.3. Industry segmentation of the South African food retail industry as percentage share by value, 2007 (Datamonitor, 2008a,b).

branded operation include a wide variety of activities and outlets; including hypermarkets, supermarkets, family franchising, mini market franchising and home shopping. Datamonitor (2008a,b) listed that Pick 'n Pay Stores has 159 supermarkets, 190 franchise stores and 16 hypermarkets. The non-Pick 'n Pay branded operations form part of the Group Enterprise Division, and includes Boxer Supermarkets, Score Supermarkets, TM Supermarkets, Property and Go Banking. Other non food retail stores include 24 clothing stores and 36 liquor stores (Pick 'n Pay, 2009).

The Shoprite group of companies was established in 1979, when eight supermarkets were purchased in the Cape Province. In 1991, Grand Bazaars was acquired, and the turnover increased significantly. Between 1992 and 1993, the Shoprite group and Checkers Supermarkets merged, to generate a combined turnover of R5.3 billion – 365.5% more than Shoprite obtained by its own in the previous year. In 1996 Sentra was acquired, and in 1998 OK Bazaars. Shoprite currently trades with 1,220 corporate outlets in 17 countries, and had total turnover of R47 billion for the financial year ending June 2008, which is 16.2% higher than the previous year. Among the different outlets included in the Shoprite holdings scope is the Shoprite Checkers Group, which include Shoprite supermarkets (373), Checkers supermarkets (119), Checkers Hyper (24), Usave stores (106), OK Furniture outlets (180), House & Home stores (35), Hungry Lion fast food outlets (113), distribution centres (20) and the OK Franchise Division, which includes OK Foods supermarkets (24), OK Grocer outlets (61), OK Mini market convenience stores (30), OK Power Express stores (14), Sentra stores (74), OK Value stores (19), Megasave wholesale stores (48) and OK Power Express stores (15).

SPAR originates from the Netherlands, where its first store was opened in 1932 under the concept of 'voluntary trading'. With the emergence of grocery chains in South Africa in the1960s, a group of eight wholesalers acquired the rights to trade under the name SPAR. By 1963 they were servicing 500 small retailers. The group expanded over time by a number of mergers and takeovers by the SPAR group Ltd., which today operates six distribution centres (in South Rand, North Rand, KwaZulu-Natal, Eastern Cape, Western Cape and the Lowveld region) and supplies goods and services to 1,422 SPAR branded stores across the country. In 1978, Tiger Brands acquired a 30% equity interest in SPAR. The company again became a wholly-owned subsidiary of Tiger Brands in 1988. In October 2004, SPAR unbundled from its holding company, Tiger Brands Limited, and listed as a company on the Johannesburg Stock Exchange (JSE). The group has showed a steady increase in total turnover since 2000. Between 2004 and 2005 the growth rate decreased slightly, but recovered well to 25% in 2006, 28% in 2007 and 23% in 2008, obtaining a turnover of R26 billion for the 2008 financial year ending 30 September 2008. The SPAR organisational structure consists of SPAR Retailers, who are independent store owners, and SPAR Distribution Centres, which provide leadership and services to the SPAR Retail members. The group has three store formats, SPAR (457 outlets) for neighbourhood shopping, SUPERSPAR (218 outlets) for one-stop, competitively priced bulk shopping and KWIKSPAR (150) for every day convenience. The SPAR Group expanded with two initiatives; 'Build It' (building materials and hardware chain) and 'Tops at SPAR' (stand-alone liquor stores).

Woolworths Holdings Limited is an investment holding company with two major trading subsidiaries, Woolworths (Proprietary) Limited, in South Africa, Africa and the Middle East, and Country Road Limited, in Australia, New Zealand and Singapore. Woolworths Food Division's financial results show very positive and stable figures. The total revenue in the food division has increased every year, reaching R10.3 billion by the financial year ending 30 June 2008, which constitutes for 51.6% of the Woolworths group – with a turnover of R20 billion in 2008. The food division only contributed 35% of the company's turnover in 2000. The Woolworths group offers a wide selection of ranges of apparel, cosmetics, toiletries, footwear, jewellery and food under its own brand name. Woolworths targets the wealthiest South African consumers, with a Living Standards Measurement (LSM) target of 6 and above. The target market of the food division is specifically LSM 9-10, though recent opportunities have been identified for expanding the target market to LSM 8 categories. The stores have a small number of branded products, and a number of private labelled products with its own Woolworths brand.

The annual turnover of each of the four major supermarket chains between 2002 and 2008 are indicated in Figure 7.4.

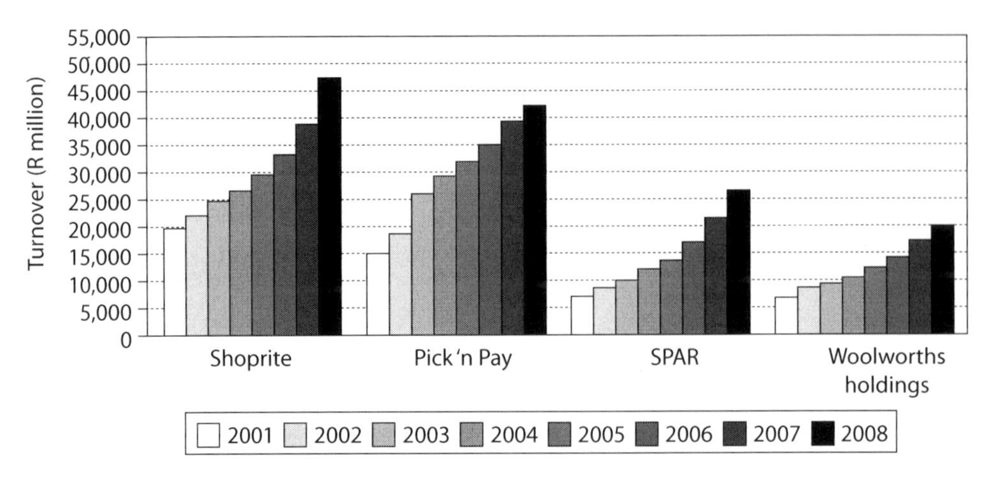

Figure 7.4. Growth in total annual turnover (R million) for each of the four leading supermarket chains between 2002 and 2008 (based on Pick 'n Pay, 2009; Shoprite, 2009; Spar, 2009 and Woolworths, 2009).

7.4 Developments in the South African food retailing environment

Reardon *et al.* (2003) conceptualised the determinants of the diffusion of supermarkets in developing countries as a changing system of demand for and supply of supermarket services. Supply incentives (motivation for supermarket expansion) and supply capacity (enabling environment promoting supermarket growth) of supermarket services, together with demand incentives (urbanisation, westernisation, etc.) and demand capacity (consumer's per capita income and access to and/or ownership of facilities and amenities) for supermarket services, sufficiently explains the changes and development of the international food retailing industry. Within a liberalised, modern market, changes in the supply and demand system determine the direction of growth of the industry. Previously, changes occurred due to new science and technology and new business management but, more recently, the demand side has played a much more important role. The development of the food retailing environment is explained in this section particularly, but not exclusively. for South Africa. Figure 7.5 provides an illustration of the determinants of change and growth in the food retail industry's system of demand for and supply of supermarket services.

7.4.1 Demand for supermarket services

The demand for supermarket services includes changing demand incentives, increasing demand capacity and evolving consumer trends. Evolving consumer trends mainly include factors and trends, such as population demographics and globalisation.

Changing consumers are characterised by adaptation of tastes and changes in various characteristics. Kinsey (1999) explains how households in general became more

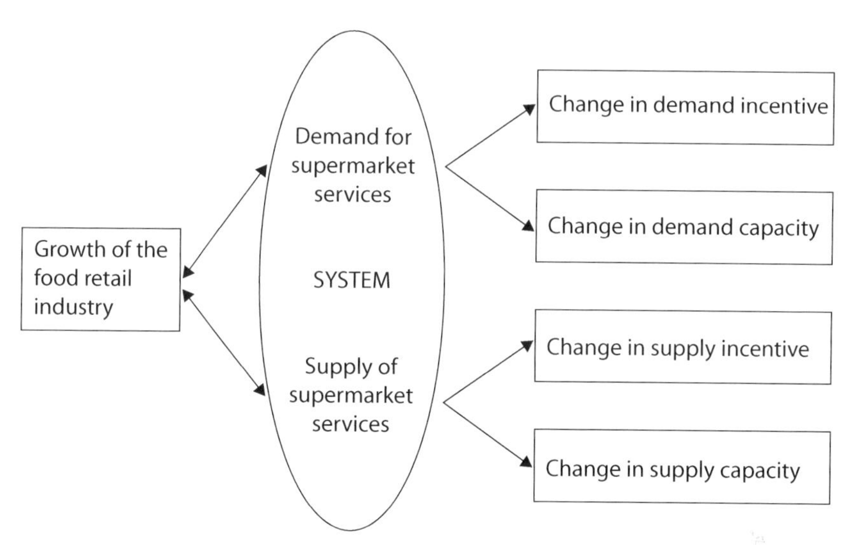

Figure 7.5. Schematic explanation of the determinants of the evolving food retail system.

heterogeneous, becoming smaller and richer, and being more likely to have a female household member in the labour force. Society in general has become more health conscious, mainly because of the rising personal cost of healthcare and the importance of the 'looking good' concept as propounded by the media. Other consumer considerations that have been brought about by information are concerns about food safety and the impact of food production on the environment. According to Hughes (2004) the trends discussed above confirm that the world's population is becoming better educated and informed, household numbers are increasing as household size decreases and increasing numbers of women participate in the labour force, resulting in dual-income households. These factors lead to a demand for more convenient food. The demand has become highly sophisticated and shifted towards added convenience and specific broadened choices. The demand for new foods, new ingredients and high taste profiles are consequences of demographic and lifestyle changes. Convenience has become a major trend, as more women enter the labour force and average household income increases (Lord, 2005). Other trends, such as the increased popularity of ethnic foods, and accentuated consciousness of health-related issues, are experienced by both urban and rural communities, throughout the world.

Changing demand incentives are characterised by the current urbanisation and general westernisation trend of the South African population. Hagen (2002) confirms that trends such as industrialisation and urbanisation in developing countries increase consumers' dependency on supermarket services. Longer working hours, diminishing leisure time, the greater role played by women in the work place and greater availability of information have had a significant influence on South Africa's food market place. Hughes (2004) explains that, on a smaller scale, increasing numbers of people in developing countries are relocating to urban areas. The main reason for doing this is their search for more and better educational and employment opportunities. The perceived and predicted movement and migration of people from rural to urban areas, from 1950 to 2020, is shown in Figure 7.6.

In the case of urbanisation, an important change in consumer demographics has been caused by a greater number of women becoming economically active. The fact that women have entered the labour force, especially in poor rural communities, has brought about significant change in households' food consumption. Kinsey (1999) states that these women have less time to shop and cook and have more money to buy more expensive ready-to-eat food, which contributes to the increased demand for convenience, more purchases of food on the go, and a decline in purchasing of raw materials for home cooking. These issues, relating to westernisation and urbanisation, determine the demand for services and are associated with a more westernised lifestyle.

It is argued that the demand capacity of South African consumers improved in most of the developing countries, due to increased real mean per capita income, on macro level, and the increase in ownership of or access to facilities and amenities, on the micro level. Senauer and Goetz (2003) state that the greatest market growth for high-value foods will be created by the sizable and rapidly growing middle class in developing countries. However, it is important to mention less sanguine effects concerning the adoption of high consumption lifestyles by rural and traditional communities in Africa and other developing regions. These involve changes in the use of technology, such as transport, refrigeration, etc. According

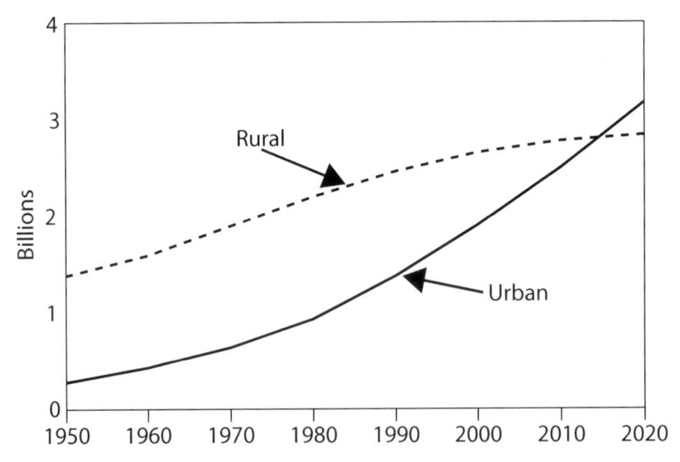

Figure 7.6. Perceived and predicted migration of people in developing countries, from rural to urban areas (1950-2020) (Hughes, 2004).

to SAARF (2002), the Living Standard Measurements (LSM)[7] in South Africa shows a drastic upward movement, with the percentage of South Africans classed as LSM 1 declining significantly from 20% in 1994 to 5% in 2001. The middle classes (LSM 4-6) have increased significantly, especially in urban areas. Figure 7.7 shows changes in the different LSM groups in South Africa over the past decade.

The increasing ownership of facilities, such as refrigerators, enables people to adapt their purchasing patterns. With such facilities, the people can shop less frequently, for they can preserve and keep food fresh for longer periods. Ownership and access to any means of transport makes it possible for customers to visit the market outlet of their choice, thereby adding to their demand capacity. The percentage of ownership of facilities in South Africa in 1994 and 2001 are shown in Table 7.2.

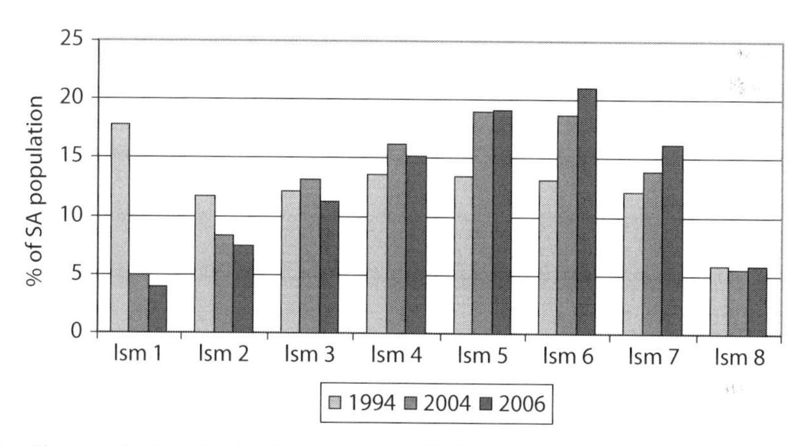

Figure 7.7. Change of living standards measurement (LSM) groups in South Africa between 1994 and 2006 (SAARF, 2007).

[7] The SAARF LSM is one type of segmentation tool based on wealth, access and geographic indicators. Variable are adjusted over time, considering its relevancy and level of luxury. The latest LSM variables included to group people with the level standard are: electric stove/hotplate, microwave oven, flush toilet in house or on plot, domestic worker, VCR in household, vacuum cleaner/floor polisher, 1 or more sedan cars, washing machine, TV set, home telephone, Hi-Fi/music centre, DVD player, built-in kitchen sink, hot running water, fridge/freezer, deep freezer, water in home/on plot, MNET/DSTV subscription, dishwasher, sewing machine, cell phone in household, PC in home, tumble dryer, two radio sets in household, metropolitan dweller, house/cluster/town house, and home security service.

Table 7.2. Percentage of ownership of food preserving and preparing facilities in South Africa (SAARF, 2002).

Food preserving and preparing facilities	% ownership in 1994	% ownership in 2001
Electricity	58	80
Fridges	48	61
Fridges (rural areas)	14	31
Electric hotplates	5	23

7.4.2 Supply of supermarket services

The supply of supermarket services is explained by supply incentives and supply capacity. Strong supply incentives are exhibited through the current motivation of supermarkets to expand by Foreign Direct Investment (FDI) and consolidation between companies, such as mergers and acquisitions (M&A). Until the 1980's supermarkets played a relatively small role in the food retail industry in Africa. In the past, only domestic capital was used, and therefore very little innovation and entrepreneurship was incorporated into supermarket development strategies. Development through FDI is the consequence of saturated local markets in developed countries, which caused intense competition. According to Farina and Dos Santos Viegas (2003) the large supermarket companies could also generate much higher margins in developing markets. This gives a multinational company a competitive advantage in the domestic market place. These large companies initially experienced little competition, with weak resistance from traditional markets and local, mostly independent, supermarkets. The supermarket companies still had to adapt to new and unfamiliar markets, especially if the market was already well developed and demand driven.

According to Reardon *et al.* (2003), consolidation in the supermarket industry has occurred in two main ways; (1) foreign acquisition of local chains and (2) absorption of smaller chains and independent stores by large domestic chains. This trend started in well-developed market places, and resulted in highly concentrated markets with intense levels of competition. Another supply incentive for progression within the supermarket industry is joint ventures within global multinationals as well as smaller companies. The main driving forces behind mergers are to achieve economies of scale, expansion of target markets, greater bargaining power with manufacturers, more efficient use of transportation and procurement systems and utilisation of information technology (Kinsey, 1998a). Companies with specific target markets can expand their markets by merging with companies with different target markets. An example of such consolidation is that of the South African company, Pick 'n Pay, which merged with a smaller local chain, Boxer. Boxer (Pty.) Ltd. targeted rural, poor communities. This merger gave Pick 'n Pay the opportunity to enter a market sector in which, at that stage, it was not active (Competition Tribunal, 2002).

Supply capacity can be explained by the enabling environment (external influences as well as self-induced development and innovative actions) towards promoting growth for supermarkets. The liberalisation of most African countries' markets was experienced as a capacitating factor for entry of supermarkets. Changes and improved political stability in various African countries also contributed to more favourable investment opportunities. The 1990's liberalising international investment policies and the appropriate timing thereof created an enabling environment for expansion of supermarkets. Changes in political conditions also contribute to change in capacity and incentive for FDI to or from certain countries.

Supermarkets also capacitated themselves to some extent by initiating a revolution regarding development of retail procurement logistics, chain integration and highly technologically advanced inventory management systems. The Competition Commission (2002) notes that, in the South African case, less than 50% of total merchandise is sold via traditional supply chains. Approximately 40% of sales bypass the wholesaler, moving directly from the manufacturer to the retailer. 5% of sales occur between the wholesaler and the consumer, and the remaining 5% occur directly between the manufacturer and the consumer. These figures indicate that the second most common wholesaler/retailer interaction in South Africa involves the self-distributing retailers, who sometimes acquire their own distribution centres. Supermarket chains usually maintain their own distribution systems, using warehouses to allocate goods to branches. The trend shows that most of the larger retailers are increasingly outsourcing the services of warehousing and transportation. The large retailers' suppliers deliver products to central depots and warehouses, from where products are distributed to retail outlets or shops. The evolution of hypermarkets involved changing the traditional distribution chain by purchasing directly from manufacturers and bypassing wholesalers. In this way they achieved high turnovers with low margins, thus placing price pressure on all competing outlets.

7.5 Impact of the rise of the supermarket industry on the agricultural sector

The rise of supermarkets in South Africa as the largest and dominant role player in the food retailing industry has affected the market place in various significant ways. The supermarket industry are very concentrated with only four large role players, whom obtain the necessary capacity and incentives to supply supermarket services wherever the demand capacity and incentive are sufficient or could be created. A highly concentrated market, such as the supermarket industry – thus the entire South African food retailing industry – are driven within the high and intense level of competition to satisfy their customers' needs. This reflects on their standards and requirements set for suppliers, especially in the agricultural sector. The impact of the rise and concentration of the supermarket industry can be explained by its effect on prices, regional economies, the traditional retail outlets, and the agrifood sector.

7.5.1 Effect on prices

A major concern relating to the effect of concentration on food retailing level is the effect on prices, which affects both the buying (suppliers) and the selling (consumers) side of retailing. According to Kinsey (2004) concentration tends to be associated with both increased and decreased prices. There are two dominant hypotheses about the relationship between market concentration and product prices. One, the mainstream school that relies on specific oligopolistic models, predicts that more concentrated markets lead directly to higher prices.

The second is the Demsetz (1973) hypothesis, which postulates that concentrated markets can experience economies of scale, lower costs and higher profits. McFall Lamm (1981) states that higher concentration is associated with lower prices, if scale economy effects outweigh the effects of collusion. Anderson (1990) suggests that, when more concentrated markets offer higher levels of service, their prices may be higher, but this is due to consumer demand for services. Weiss (1989) and Schmalensee (1989), as cited by Kinsey (1998b), reviewed various studies done on concentration and prices. Weiss found concentration and prices to be positively correlated in 73% of reviewed cases. Schmalensee concludes with strong evidence of concentration raising prices.

Furthermore, concentration at the wholesale level may lower food prices. Most studies conclude that lower costs are made possible by economies of scale in procurement, vertical coordination with suppliers and better use of information technology. Anderson (1993), as cited by Kinsey (2004), points out that, in concentrated markets where large supermarkets have correspondingly large market shares, the supermarkets will offer lower prices because they have real economies of scale and they will compete for customers by lowering prices.

According to Connor (1994), the oligopsonistic and oligopolistic powers of retailers have been known for their influence on price stability. He also states that it is a well-known principle of economics that firms with market power display more price rigidity over time than powerless firms. As food retailers' market power increased over the past decade, they became less inclined to change the prices of individual products in response to changes in agricultural prices and general consumer food inflation changes. The Competition Commission (2002) states that, although retailers' costs are invisible, retailers' prices are completely visible. Transparencies facilitate inter-firm monitoring, but sometimes also make it possible to spot collusive price fixing.

King *et al.* (2004) hypothesise higher prices, specifically in low-income areas, by reasoning that operating costs are higher for stores that serve low-income households. They found that supermarket operating costs may be higher in low-income areas due to higher occupancy costs, less efficient store designs, higher rates of labour turnover, and/or greater losses due to theft. Smaller average transaction sizes and procurement inefficiencies due to smaller orders to suppliers may also increase costs.

7.5.2 Effect on regional economies and traditional retail outlets

The rapid spread of supermarkets is driving many traditional food retailers, such as small corner stores, out of business and taking business from others, such as street markets (Reardon *et al.*, 2003). According to Capps and Griffin (1998), the larger sized retail markets, such as supercenters and hypermarkets, are the prime retail growth vehicles of the future. It is interesting to speculate what effect a highly concentrated and growing supermarket industry would have on regional economies and traditional retail outlets, taking into consideration cost effectiveness, the change in prices and the centralisation trend.

Woo *et al.* (2001) found that traditional food retail outlets face serious competition from supermarkets mainly because of their low-price appeal to consumers. Martens *et al.* (2005) state that shoppers patronise traditional supermarkets less often as they shift some of their purchases to supermarkets, and that this shift is moving sales from small markets to larger ones and forcing small grocers to close. They conclude that, in general, businesses competing with large supermarkets will be harmed significantly, but those in the region who do not compete directly, will benefit. Roe *et al.* (2005) found that food purchased in supermarkets is not a perfect substitute for food purchased in traditional outlets, because, as income increases (such as the per capita income growth currently experienced in South Africa), the total share of income spent on food will decline, but the share spent in traditional markets will decline more rapidly than the share spent in supermarkets.

Some researchers suggest survival strategies for the traditional food retail sector. Woo *et al.* (2001) noted that upmarket outlets, which target higher income, niche market segments, were found to be less affected by supermarkets and able to continue in the market in spite of relatively higher prices. Capps and Griffin (1998) found that, in order to stabilise market share, traditional outlets must be fully aware of and conscious about changing consumer trends, needs and preferences, especially in an effort to reduce prices and change labour utilisation, and yet maintain profit margins.

Martens *et al.* (2005) raise the concern that large food retailers are using their lower cost structure and advantages in marketing, store design and shelf space allocations to reduce consumer access to local small retailers, increase retailer market power and discourage competition. However King *et al.* (2004) state that the introduction of successful supermarkets can encourage low-income areas to become more vital and viable, not only by providing more shopping opportunities in a more competitive environment, but also by creating new employment opportunities for area residents.

Major retailers have achieved growth by franchising and buying out smaller independent retailers, instead of establishing large format stores. The number of grocery and convenience stores has increased over the past few years, and many grocers have joined franchise operations. As the number of outlets in a given supermarket chain increases, the tendency is a move from a per-store procurement system, to a distribution centre serving several stores in a given zone, district or region. This is accompanied by fewer procurement units and increased use of centralised warehouses. Centralisation increases efficiency of procurement by reducing coordination and other transaction costs, although it may increase transport costs by extra movement of the actual product (Reardon *et al.*, 2004). This trend of centralisation results in a decrease in procurement from and support of regional/local economies, through local agricultural producers, local suppliers, local institutions and therefore local consumers. Money generated by regional economies can decrease drastically, due to further multiplier effects of decreasing local support, regardless of benefits to the consumer tends in the form of lower consumer prices as a result of lower transaction costs.

Since the 1990's the South African food retail sector has experienced a steady contraction of small retailers, mergers and acquisition among retailers, and the emergence of new national retailers. According to Martens *et al.* (2005) the grocery industry shifted from an industry dominated by small grocers serving local markets, to one characterised by large retailers operating in international markets. The growth of large grocers is explained by economies of size and scope, the adoption of advanced information technology, supply chain management strategies, which drastically lower their costs, fewer weekly trips to supermarkets by consumers, and by evolving store formats. Supermarkets, as mass merchandisers in the South African context, have a dual objective – qualitative (to increase quality and eventually safety of the product) and quantitative (to reduce costs and increase volumes procured) (Reardon *et al.*, 2004). This provides supermarkets with a competitive advantage over smaller, traditional retailers. In conclusion, with regard to the supermarkets' effect on regional economies and traditional retail outlets, consider that, since the traditional food production – marketing – retail outlets are relatively small (in sales and scope) and labour intensive, and since traditional suppliers tends to lack the advantages of large, centralised suppliers, capital deepening and a very competitive market place leads to an expansion of the large, scale-efficient food channels, while the traditional retail system experiences a departure of labour and little, if any, growth in the national and regional economy. Another recent addition to the competition framework is recognised by Binkley and Connor (1996), who conclude that the degree of supermarket rivalry is no longer the only important competitive force, and that competition from new store formats (hypermarkets and convenience stores), fast food outlets and small niche sellers can affect sales and prices of supermarkets.

7.5.3 Effect on the agrifood sector

Organisational change, accompanied by intense competition in the food retailing sector, has driven changes in the organisation of procurement systems of supermarket chains towards centralised and regionalised distribution, use of specialised/dedicated wholesalers and preferred supplier systems and demanding private quality standards. The effect of supermarkets on the agrifood sector relates mostly to greater demand for agricultural products and access for primary producers to sell to supermarkets. Meeting the requirements caused by organisational change in supermarket procurement systems present clear challenges and opportunities for producers. Humphrey (2007) depicted that as the retail environment in Africa transforms, increased small farmer output could fail to gain access to the most dynamic and expanding marketing channels because it does not have the quality assurance, volume or traceability that a modern retail sector requires. Cacho (2003) also recognises the smallholder farmers, in particular, are at risk of being marginalised further by the swift introduction of a 'new' market that has specific requirements.

The cartel-like framework of the buyer's side can classify the producers as 'winning' farmers and 'loser' farmers. Cacho (2003) concludes that large-scale farmers – at national or regional levels – who 'self select' by complying with supermarket 'market' requirements, would meet this challenge and become the 'winning farmers'. In the 'loser column' of this comparison framework are the smallholder farmers who do not and/or cannot 'self select'. Reardon *et al.* (2004) state that the specific requirements farmers have to comply with will vary by type of retailer, and include factors such as size of the retail chain or consumer segment targeted. These factors then translate into requirements for product quality and post-harvest practices by producers and others actors in the supply chain. Reardon *et al.* (2004) state that adopting new practices can open doors to suppliers, enabling them to sell to supermarket chains that are 'growing' the market in terms of volume, value adding and diversity. A supplier can move from being a local supplier to being a national, regional or global supplier – and therefore categorise himself as a 'winning farmer'. On the other hand, Roe *et al.* (2005) acknowledge that the rapid adjustment in the food marketing chain associated with the growth of supermarkets has raised concern about the plight of smaller, traditional farmers who cannot meet the more demanding market channel standards, and therefore become the 'loser farmers'. These farmers will typically supply local, more traditional retail outlets.

Reardon *et al.* (2004) characterises the determinants of a producer's 'channel choice' as the products and transaction attributes required and price given by the buyer in a given channel, for a specific product of a particular variety and grade. Important to the producer is, furthermore, costs and benefits to small and medium agro-food entrepreneurs involved in entering the supply channel for supermarkets. It is important, in this regard, to understand technology options that deliver the same attribute set and quantity – in economic terms, the 'isoquants' in factor space. That refers to the required set of attributes that is delivered by producers using capital-using versus capital-saving technologies. Reardon *et al.* (2004) believe the analyses relating to implementation of the conceptual framework involve two stages, firstly involving the farmer selecting the market channel according to his incentives and capacities, and secondly selecting the technology, again as a function of his incentives and capacities, embodied in a set of adoption choices of capital items.

There are certain noteworthy benefits for the 'winning farmers', such as higher prices and more markets. Empirical studies of farm-to-retail price transmission have found asymmetry in price responses; retail prices often do not fall to the same extent as farm level prices, and prices rise less than proportionately to farm level price increases. Backward integration of retailers increases the likelihood of direct buying from agricultural producers. Price rigidity, price evaluation and price asymmetry all contribute to reductions in producer welfare and suggest a need for better antitrust enforcement for agriculture (Carstensen, 2000). Reardon *et al.* (2004) cite a statement by Javier Gallegos, head of marketing for Hortifruti:

> The market is fragmented, unformatted, unstandardised. The growers produce low quality products, use bad harvest techniques, there is a lack of equipment and transportation, there is deficient post-harvest control and infrastructure, and there is no market information. There are high import barriers and corruption. The informal market does not have: research, statistics, market information, standardised products, quality control, technical assistance and infrastructure.

Reardon *et al.* (2004) state that, in order to close the gap between supplies and needs, supermarket chains in developing regions have been moving away from the old procurement model, based on sourcing products from the traditional wholesaler and the wholesale markets, towards the use of four key pillars of a new kind of procurement system:
1. 'specialised/dedicated wholesalers and suppliers';
2. centralised procurement through distribution centers;
3. assured and consistent supply through 'preferred suppliers'; and
4. high quality and increasingly safe products through private standards imposed on suppliers.

Farmers are obliged to adapt if they wish to be included in the new procurement system. Direct contracting has been, to date, the best way of entering this market. Humphrey (2007) warns that the switch from small farmer sourcing depends upon the availability of produce from alternative sources (large farms and imports), the cost effectiveness of outgrower schemes and the political expediency of working with small farmers. Reardon *et al.* (2004) state that, to enter this market, producers sometimes need to make rather large and significant investments in human capital, management, input quality and basic equipment. Roe *et al.* (2005) express concerns about the possible effect of supermarket expansion on the welfare of traditional farmers, especially regarding the relative capital intensity of modern marketing channels compared to traditional channels. The supermarket industry is caught up in this highly competitive environment, and therefore seeks to lower product and transaction costs and risk. This will secure the position of more capable farmers. It is inevitable that standards demanded by consumers will increasingly be a major driver of concentration in the farm sector in developing regions. Sooner or later, retail concentration will cascade into supplier concentration.

Recommendations to small farmers to enter the market are as follows:
- Establishment of partnerships between small farmers, with inadequate resources, and financiers, insurers, extension services and supermarkets, to assist in their lacking needs.

- Assistance programmes to small farmers to enable them to engage with training institutions and agribusinesses and/or supermarket groups to enhance their technical and managerial capacity.
- Assistance programmes to small farmers to enable them to approach supermarkets with their business plans to establish coherence with the supermarkets' enterprise development projects and/or corporate social investment initiatives.
- Collective financing, production and/or marketing amongst small farmers, to ensure sufficient volume to justify the contract and to enable them to supply more continuously.
- Utilisation of Fairtrade, or related branding of products obtained from small and emerging farmers.
- Direct contracting between farmers/group of farmers and the supermarket, or its procurement company.
- Establish outgrower schemes facilitated by supermarket procurers, to ensure efficiency in production and marketing.
- Obtain information on improving production techniques and quality to ensure compliance with required standards.

7.6 Conclusion

South Africa has a very dynamic food retail industry, defined by an intense level of competition, concentration, and a very unique South African (and nowadays African) customer base, which vary from poor in rural areas to the rich in the cities. The South African food retail industry is dominated by the supermarket industry, which controls the larger part of the industry, together smaller chains and independent supermarkets, as well as an informal market. The spread of supermarkets are determined by the supply incentives and capacity for supermarket services. However, supermarkets need to take careful consideration of the demand incentives of different customers and reflect on the demand capacity, especially when considering location and target market. However, the food retailing industry remains very demanding and needs to adapt according to changing consumer trends and to improve logistical and other management aspects to maintain profit margins.

The impact of the developing food retailing industry and the concentrated supermarket industry reflects on different levels of the industry, because a hegemonic system depends on an oligopolistic market structure to survive, because too many competitors dilute control over suppliers, and a monopoly reduces incentives to innovate and improve performance. An important underlying question remains, 'How does the food retail industry, mostly referring to the supermarket industry, affect the market place?' Does it lead to higher or lower food prices, better or worse services, more or fewer choices between stores and among products, and more or fewer employment and earning opportunities in the agri-food sector? The argument surrounding the effect of the expansion of supermarkets in South Africa tends to be very contradictory. The introduction of dynamic food retailers, such as supermarkets, in Africa can have either positive or negative effects. It could positively influence the area by providing local people with access to good quality produce, convenience shopping and lower prices for basic household goods. According to Weatherspoon and Reardon (2003), the rural poor need food markets to escape from poverty. It is therefore inevitable for food policies to be based on a sound understanding of the way the changing food marketing system works, especially for the rural and poor population. The negative aspects include issues such as the exclusion of traditional markets by specialised retailers, reduction of market access possibilities for local and small-scale producers and suppliers, and limiting the potential for money generation within the region. Each region experiences different effects on the regional economy, traditional food retailing and agri-food sectors. It seems, however, that consumer welfare with respect to food and general economic growth have benefited from the expansion of supermarkets into Africa. The impact on the agricultural sector depends on the agricultural sector themselves. The supermarket industry has been criticised for their inability to negotiate fair prices for farmers, however, no recognition is being taken that they do not have negotiation power when it comes to unsatisfied customers. This reflects on high set standards and requirements by the supermarket procurers to their suppliers. Farmers, who adapted accordingly to enable themselves to provide quality products on a continuous basis at the required volumes, are gaining significantly with a secure (contracts) market and prices. Accessibility of this market for small farmers, depend on their ability to produce good quality products and possibly collective supply and good structure organisation among themselves, and possibly the supermarkets themselves.

References

AC Nielson, 2004. Average turnover, and store numbers. Retail store data. AC Nielson, Ormonde, South Africa.

Anderson, K.B., 1993. Structure-performance studies in grocery retailing: a review. In: R.W. Cotterrill (ed.) Competitive strategy analyses in the food system. Westview Press, Boulder, CO, USA, pp. 201-219.

Binkley, J.K. and J.M. Connor, 1996. Market competition and Metropolitan-area grocery prices. Private strategies, public policies & food systems performance. Working paper #44. Purdue University, West Lafayette, IN, USA. Available at: http://ageconsearch.umn.edu/handle/25988.

Business Day, 2002. Score sells Tanzanian stores to Shoprite. Business Day, 25 November 2002. Available at: www.liquidafrica.com.

Carstensen, P.C., 2000. Concentration and the destruction of competition in agricultural markets: the case for change in public policy. Wisconsin Law Review 2000: 525-537.

Cacho, J., 2003. The supermarket 'market' phenomenon in developing countries: implications for stallholder farmers and investments. American Journal for Agricultural Economics 85: 1162-1163.

Capps, O. and J.M. Griffin, 1998. Effect of a mass merchandiser on traditional food retailers. Journal of Food Distribution Research 29: 1-7

Competition Commission, 2002. Inquiry food price rises, final report. Available at: http://www.compcom.co.za/policyresearch/policyresearch.asp.

Competition Tribunal, 2002. Government of the Republic of South Africa Competition Tribunal: Case no 52/LM/Jul02: In the large merger between Pick 'n Pay Retailers (Pty) Ltd. and Boxer Holdings (Pty) Ltd. Available at: http://www.comptrib.co.za/assets/uploads/Case-Documents/52LMJUL02.pdf.

Connor, J.M., 1994. North America as a precursor of changes in Western European food-purchasing patterns. European Review of Agricultural Economics 21: 155-173.

Cotterill, R.W., 1986. Market power in the retail food industry: evidence from Vermont. The Review of Economics and Statistics 68: 379-386.

Datamonitor, 2008. Food retail in South Africa: industry profile. Available at: www.datamonitor.com.

Datamonitor, 2008. Global food retail: industry profiles. Available at: www.datamonitor.com.

Demsetz, H., 1973. Industry structure, market rivalry, and public policy. The Journal of Law and Economics 16: 1-9.

Farina, E.M.M.Q and C.A. Dos Santos Viegas, 2003. Foreign direct investment and the Brazilian food industry in the 90's. International Food and Agribusiness Management Review 5 (2). Available at: https://www.ifama.org/publications/journal/vol5/cmsdocs/efarina.PDF.

Goldman, A., S. Ramaswami and R.E. Krider, 2002. Barriers to the advancement of modern food retail formats: theory and measurement. Journal of Retailing 78: 281-295.

Hagen, J.M., 2002. Causes and consequences of food retailing innovation in developing countries: supermarkets in Vietnam. Paper presented at the Annual Meeting of the Western Coordinating committee #72. June 2002, Las Vegas, USA. Available at: http://EconPapers.repec.org/RePEc:ags:wccstw:16612.

Hughes, D., 2004. Consumption trends. AMT/Landbouweekblad/Vleissentraal Seminaar: Castle Kyalami, Pretoria, South Africa.

Humphrey, J., 2007. The supermarket revolution in developing countries: tidal wave or tough competitive struggle? Journal of Economic Geography 7: 433-450.

King, R.P., E. Leibtag and A.S. Behl, 2004. Supermarket characteristics and operating costs in low-income areas. Agricultural Economic Report No. 839. USDA, USA. Available at: www.ers.usda.gov/publications/aer839/aer839.pdf.

Kinsey, J.D., 1998a. Supermarket trends and changes in retail food delivery. Agricultural Outlook Forum, February 24, 1998.

Kinsey, J.D., 1998b. Concentration of ownership in food retailing: a review of the evidence about consumer impact. Working paper, The Food Retail Industry Centre, University of Minnesota, St. Paul, MN, USA. Available at: http://purl.umn.edu/14329.

Kinsey, J.D., 1999. The big shift from a food supply to a food demand chain. Retail Food Industry Center, University of Minnesota, St. Paul, MN, USA. Available at: http://EconPapers.repec.org/RePEc:ags:umaema:13187.

Kinsey, J., 2004. Emerging trends in the new food economy: consumers, firms and science. Paper presented at the International Agricultural Trade Research Consortium (IATRC). Annual Meeting, December 15-17, Monterey, CA, USA. Available at: http://purl.umn.edu/14575.

Lord, J.B., 2005. How the changing consumer drives food retailing strategy and structure. Presentation at the Saint Joseph's University, Philadelphia, PA, USA.

Martens, B., R.J.G.M. Florax and F. Dooley, 2005. The effect of entry by supercenter and warehouse club retailers on grocery sales and small supermarkets: a spatial analysis. Paper presented at the American Agricultural Economics Association Annual Meeting, Providence, Rhode Island, July 24-27, 2005.

McFall Lamm, R., 1981. Prices and concentration in the food retailing industry. The Journal of Industrial Economics 30: 67-78.

Pick 'n Pay, 2009. Annual report June 2008. Johannesburg, South Africa. Available at: www.picknpay.co.za.

Reardon, T. and J.A. Berdegue, 2002. The rapid rise of supermarkets in Latin America: challenges and opportunities for development. Development Policy Review 20: 371-388.

Reardon, T., C.P. Timmer Barret and J.A. Berdegue, 2003. The rise of supermarkets in Africa, Asia, and Latin America. American Journal of Agricultural Economics 85: 1140-1146.

Reardon, T.A., J.A. Berdegue, M. Lundy, P. Schutz, F. Balsevich, R. Hernandez, E. Perez, P. Jano and H. Wang, 2004. Supermarkets and rural livelihoods: a research method. Staff paper, Michigan State University, East Lansing, MI, USA. Available at: http://EconPapers.repec.org/RePEc:ags:midasp:11818.

RocSearch, 2004. South Africa: food retail industry. RocSearch Limited, London, UK. May, 2004.

Roe, T., M. Shane and A. Somwaru, 2005. The rapid expansion of the modern retail food marketing in emerging market economies: implications to foreign trade and structural change in agriculture. Paper presented at the American Agricultural Economics Association Annual meeting, Providence, Rhode Island, July 24-27, 2005. Available at: http://purl.umn.edu/19112.

SAARF, 2002. Big improvement in South Africans living standards post 1994. South African Advertising Research Foundation (SAARF), Bryanston, South Africa.

Schmalensee, R., 1989. Inter-industry studies of structure and performance. In: R. Schmalensee and R. Willig (eds.) Handbook of industrial organization. Elsevier, New York, NY, USA, pp. 951-1009.

Senauer, B. and L. Goetz, 2003. The growing middle class in developing countries and the market for high-value food products. Working paper, The Food Industry Center, University of Minnesota, St. Paul, MN, USA. Available at: http://purl.umn.edu/14331.

Senauer, B. and L. Venturini, 2005. The globalization of food systems: a conceptual framework and empirical patterns. Working paper, The food industry center, University of Minnesota, St. Paul, MN, USA. Available at: http://EconPapers.repec.org/RePEc:ags:umrfwp:14304.

Shoprite, 2004. 24 year review: financial statistics for Shoprite Holdings Limited. Available at: www.shoprite.co.za.

Shoprite, 2009. Shoprite holdings annual financial report 2008. Available at: www.shoprite.co.za.

Spar, 2009. The Spar group limited Annual Report 2008. Available at: http://www.spar.co.za/.

Trail, W.B., 2006. The rapid rise of supermarkets? Development Policy Review 24: 163-174.

Wahome, M., 2001. Giant supermarkets nattle for supremacy. Business Week supplement to Daily Nation (Kenya), 11 December 2001.

Wandera, N., 2002. Uchumi set to open branch in Kampala. *East African Standard*, 13 June 2002.

Weatherspoon, D. and T. Reardon, 2003. The rise of supermarkets in Africa: implications for agrifood systems and the rural poor. Development Policy Review 21: 333-355.

Weiss, L. (ed.), 1989. Concentration and price. MIT Press, Cambridge, MA, USA.

Wilson, M., 2003. South African retail industry. Retail and consumer products. Ernst & Young. Available at: www.ey.com/za.

Winter-Nelson, A. 2002. Global supermarkets and local farmers in Tanzania. Discussion paper. Department of Agricultural and Consumer Economics, University of Illinois, Urbana, IL, USA.

Woo, B., C.L. Huang, J.E. Epperson and B. Cude, 2001. Effect of a new Wal-Mart supercenter on local retail food prices. Journal of Food Distribution Research 32: 173-181.

Woolworths, 2009. Audited group financial results, 30 June 2008. Available at: www.woolworthsholdings.co.za.

8. Unlocking credit markets

Jan A. Groenewald and Andries J. Jordaan

8.1 Introduction

Small farmers and other small entrepreneurs have internationally had to cope with serious problems concerning access to finance for their operations. And without finance, it is not possible either to produce surplus products or to market these products; commercial agricultural production and trade, immaterial of how large or small the scale, is impossible without finance.

Many rural (and also urban) households in many parts of the world have lived in a vicious cycle of poverty: low rates of capital investment, coupled with low incomes, inevitably led to low savings, low productivity and ultimately again poverty and low investment. This cycle cannot be broken without more effective functioning of the financial market, including credit institutions. The main functions of this market are (Heidhues, 1994):

- the generation and/or increase of incomes through investment and production/ marketing credit;
- stabilisation of income and consumption through savings/dissavings and consumption credit; and
- security of income by providing potential access to finance.

8.2 The supply-led approach to credit

Poorer communities, both urban and rural, have traditionally been unable to access sufficient credit. Because of a lack of collateral, formal credit institutions such as banks have not been willing to supply credit to poor entrepreneurs or SMME's. Local traders have been the main source of finance for small farmers with prospective crops being an important form of collateral. Interest rates have been perceived to be exorbitant. It was widely assumed that poor rural households cannot or will not save, because of their low incomes and a high propensity to spend their income on ceremonial activities in addition to the necessities of life. Spio *et al.* (1995) point out that this notion has internationally been proven to be erroneous. They point at research findings of rural poor people in Africa and Asia having high propensities to save and high savings rates when saving is possible. They cite research by Adams (1978), Hofmann (1990) and Ong *et al.* (1976). Similar results were obtained in South Africa by Coetzee (1988), Kuhn *et al.* (2000) and Spio (1995, 2003).

Governments in many countries, including South Africa, stepped in to close the perceived gap. This led to what has later been described as the traditional, or conventional, or supply-led approach. According to Bouman and Hospes (1994), the rural landscape in many developing countries became dotted with rural banks, co-operatives and specialised farm credit institutions. Billions of dollars were poured into a multitude of projects that claimed to be concerned with the welfare of rural households. Subsidised credit was in vogue with these schemes, in which very little, if any collateral was asked from borrowers.

Donors and policy makers thus embarked on supply-led credit and simultaneously neglected the mobilisation of rural savings. Results have practically everywhere been futile (Adams and Vogel, 1984). These programmes were characterised by limited investment in productive inputs, high default rates (up to 40% in South Africa), lack of savings mobilisation and limited client outreach (Kuhn *et al.,* 2000).

The main flaws of the supply-led approach have been summarised as follows (Spio and Groenewald, 1998; Von Pischke, 1991):
1. The overemphasis on credit led to high rates of delinquency and very poor loan recovery; internationally, loan failures averaged around 25%, rising to 80% in some cases (Ellis, 1994).
2. The supply-led approach ignores savings mobilisation, and thus reduces financial intermediation; in pursuit of this approach, many financial institutions do not also provide deposit facilities for savers. The low interest rate regime that dominated the supply-led approach was moreover a deterrent to savings.
3. This approach also ignored the importance of other necessary conditions for productive investment – e.g. good infrastructure, security in land tenure, legal protection of property rights and an economic framework within which private entrepreneurs can freely introduce new technologies.
4. The supply-led approach ignores the importance of equity finance and therefore misses out on its advantages.
5. Governments use lender targets and credit quotas to push intermediaries through the financial frontier at a faster pace than would otherwise be the case. These regulations do not address those problems that make lenders reluctant to approach the frontier voluntarily, as they are typically designed without reference to and in disregard of the cost of implementation. They result in lower resource mobilisation and they reduce the flexibility of banks to undertake discretionary lending; the state action reduces the scope for financial institutions to base the granting of credit on economic and financial criteria.
6. The approach ignores transaction costs. Although these policies were designed to benefit individual small farmers and SMME's, the approach paradoxically increases transaction costs both for borrowers and savers as lenders become less eager to award loans and demand more information. As institutions do not depend on deposits for funds, they do not seek new clients, and transaction costs become higher for depositors. Lenders' transaction costs also increase. The approach is a real cost to society.
7. Another flaw of the supply-led policies has been an overemphasis on institutions rather than on instruments in their bid to advance credit to farmers and the poor. Governments and donors have often been much more interested in the establishment

of credit institutions than in the management thereof or in the training of personnel. The result has in many cases been management failure.

The supply-led approach has clearly led to many failures. Awareness of the flaws and failures has given rise to new, more rational approaches. One should however always remember the weaknesses of the supply-led approach as political groupings do at times move for the institution of similar steps.

8.3 The demand-led approach

The failure of the supply-led approach became evident as a large number of the institutions involved proved to be non-viable and non-sustainable; many donors removed or reduced support, and governments in some developing countries could not continue institutions. Commercial banks had in some cases already withdrawn from rural areas. This meant that rural SMME's were left stranded with very little access to finance, even for the simplest of transactions. A more rational approach, the demand-led approach, gradually took over as its successes became evident.

In contrast to the supply-led approach, the demand-led approach is mainly driven by the following considerations:
1. Savings mobilisation. It is recognised that a strong savings base reduces the reliance on external funding, and that for a system to be sustainable, borrowers must also be savers. This also reduces information costs, and default rates become smaller.
2. For a finance institution to be independent, viable and sustainable, its income (interest on loans) must cover all costs, including interest paid to depositors, costs of doing transactions and risk (e.g. of default).
3. Institutional innovation. Successful commercial operation in a dynamic, developing world will, as far as finance is concerned, require innovative ways of reducing transaction costs for the institution as well as its clients, borrowers and depositors. This will most likely mean different services and service approaches in contrast to what happens in state credit entities. Viability, self-sufficiency, access and efficiency are key elements (Padmanabhan, 1988, cited by Bahta, 2003).

The demand-led approach has been characterised with much less direct government (and donor) involvement. Although many governments still support credit institutions, directly government-run institutions are now, in the 2000's, a much rarer phenomenon than was the case in the 1970's and 1980's. There now appears to be more of a realisation that governments are usually not very efficient in performing business activities.

Mobilisation of savings is an important, integral part of the demand-led approach. Spio *et al.* (1995) summarised the most important arguments for savings mobilisation, as enunciated by economists like Fernando (1991), Meyer (1989) and Vogel (1984), as follows:

1. Savings mobilisation can give the poor access to assets with higher returns than those from tangible assets (e.g. cattle kept as a hedge against hard times), and this can lead to a more equitable distribution of income. Positive real rates of interest and low transaction costs are necessary prerequisites for this.
2. Savings mobilisation enables financial institutions to improve their financial viability, liquidity, solvency and thus overall performance. Repayment performance may be better. This can help to stabilise rural financial markets and reduce their dependency on governments and donors' funds (Schrieder and Heidhues, 1991).
3. Savings mobilisation aids in reducing asymmetric information and incentive problems, thus improving relationships between financial institutions and their clients. It improves the intermediaries' ability to judge clients' creditworthiness.
4. It allows small farmers and SMME's to improve their creditworthiness through accumulation of financial assets.

A review of empirical studies conducted in a few Asian and African countries, including South Africa, has revealed considerable savings capacities, and relatively high marginal propensities to save among the rural poor (Spio and Groenewald, 1998).

8.4 The Grameen Bank

An early development concerning the demand-led approach occurred in Bangladesh. In 1976, Prof. Muhammad Yunus of the University of Chittagong launched a research project to examine the potential of designing a credit delivery system that could provide credit services to the rural poor. He started by making small loans of US $27 each to a group of 42 families in rural villages close to Chittagong so that they could create small items for sale without the burdens of predatory lending (Anonymous, 2008). The Grameen Bank process was thus started. The process was very successful, and in 1983, the Grameen bank became an independent bank. In 1991, its disbursements amounted to US$ 74.35 million; this increased to US$ 608.79 million in 2005. Total deposits were US$ 25.86 million and US$ 481.22 million in these two years respectively; women constitute 97% of its clients (Grameen Bank, 2008). The organisation and Prof. Yunus were jointly awarded the Nobel Peace Prize in 2006.

The Grameen Bank is best known for its system of group lending, also called solidarity lending. This concept has become a cornerstone of micro-credit lending in over 43 countries (Anonymous, 2008). Each member must belong to a group. Experience has shown that small groups generally do better, and that better results are obtained if groups are homogeneous. The Grameen Bank customers form five-member groups. In many instances, groups have to guarantee the pay-back of loans of their individual members collectively. This provides collateral, and peer pressure acts to ensure individuals' adherence to the rules. The Grameen bank itself does not practice joint liability, but in practice group members often contribute the defaulted amount with the intention of collecting the money from the defaulted member at a later time. This behaviour is facilitated by the Grameen Bank's policy of not lending any further credit to a group in which a member defaults. Grameen Bank has experienced very high repayment rates – over 98%, with the exception of 2001, when drought caused many loans to become overdue.

As already mentioned the Grameen bank model, or variations thereof, has been adapted in many countries. These will further be referred to as microcredit schemes, much of which is informal in nature.

8.5 Informal financial markets (microcredit)

Various types of less formal types of financial institutions co-exist in addition to the formal types such as banks. It is strictly speaking erroneous to classify these all as 'informal'; various types have also adopted types of formalism, ranging from completely informal groups with no written rules or agreements to specific legal recognition to different kinds of associations which vary in their degree of formalism. Yunus (2008) classifies microcredit into ten different groups or types. Many of these institutions fall outside the definition supplied by Chandavarkar (1988) as 'all legal but officially unrecorded and unregulated financial activities and transactions which are outside the orbit of officially regulated institutional finance'. This category stretches from moneylenders who extend credit at very high rates without any written agreements on the one hand, to more sophisticated forms which are still outside the ambit of the formal banking sector.

The informal financial sector institutions are thus very heterogeneous; they mostly function outside central bank and/or government regulation and some even have no legal standing. Whilst some have a long history and continue to operate and even thrive when the formal sector develops or expands, others emerge as reaction to the repression of the formal system, whilst some loose their importance, and pine away as financial markets are liberalised (Bahta, 2003).

Rotating Savings and Credit Associations (ROSCAs) form the predominant form of informal financial institutions all over the developing world. These are groups of individuals who have formed associations to save, share risks and borrow money. The Grameen Bank really started off as a type of ROSCA, and became more formalised with growth; today, it can perhaps best be described as an association of ROSCAs. ROSCAs are found in many countries and are known by a variety of names. Some African examples are: 'Ikub' or 'Ider' (Ethiopia and Eritrea), 'Dyanggy' (Cameroon), 'Chilemba' (Uganda, Malawi, Zambia and Zimbabwe), 'Mchezo' or 'Upatu' (Tanzania) and 'Stokvels' (South Africa) (Bahta, 2003). These are grass level groups which operate with rules, regulations and operational methods suited to local traditions and conditions. All ROSCAs have saving as a core feature. Members borrow from the credit fund for a variety of purposes, which may include both domestic and business activities (Miracle *et al.*, 1980, quoted in Bahta, 2003). ROSCAs probably include most types of microcredit as identified by Yunus (2008).

Meyer and Nagarajan (1992) and also Spio (1994) list some attributes of micro, or informal finance:

1. *Heterogeneity*. Informal finance includes a wide variety of institutional forms, and a variety of contracts between lenders and borrowers exists within each type.
2. *Specialisation*. Some groups provide only one or a very limited number of financial services, limited to clients of whom they have good information.
3. *Collateral*. Very few loans involve collateral in the classic sense. However, substitutes are used for collateral:
   a. Third party guarantees: a third person undertakes surety for the loan.
   b. Linked market or tied types: the loan is given on the promise or agreement that the lender will be the sole buyer of the product at a certain implicit rate of interest.
   c. The threat of loss of future borrowing opportunities.
   d. Group credit: borrowers belong to a group, and the group is either liable for repayment, or liable to loose their future borrowing possibilities; moral pressure within groups is a powerful form of collateral substitute.
4. *Interest rates and transaction costs*. Interest rates vary from being exorbitantly high in the case of some moneylenders whereas those of some others are much lower; friends and relatives sometimes charge no interest at all. According to Fry (1988), informal finance is normally characterised by higher interest rates than formal finance. However, non-interest costs of borrowing from banks are often substantial (e.g. transaction costs); these costs are often virtually non-existent in informal financial markets. There is a minimum of formal procedures (Meyer and Nagarajan, 1992) and also better knowledge concerning clients (thus, smaller risks) on the part of lenders, and thus low or no information costs. As interest rates are not regulated in this market, they can readily adjust to market forces; neither is informal finance subject to the reserve requirements imposed on formal institutions (Fry, 1988).

It must be pointed out that informal, or microfinance cannot cure all the credit problems of agricultural SMME's in less developed areas. There is a need for capital from outside, provided the credit practices are economically and financially rational. Governments can, and should aid in this process through specialised agencies and through the formal private banking system. Banks have also recognised the potential of expanding business and creating goodwill by forms of cooperation with the less formal microfinance system. However, forms of microfinance will for long have a crucial role to play in the development of agricultural SMME's in developing agriculture. This includes the emerging agricultural sector in South Africa.

8.6 Alternative finance sources in South Africa

The transformation of the South African financial sector since 1994 is in line with global trends and will continue, bringing with it potential opportunities such as the roll-out of the second wave of Public Private Partnership (PPP) projects; the further development of e-commerce, and the expansion of the micro-lending business – servicing the needs of the 'unbanked' majority of the population.

The South African government realised the importance of rural finance and the reports from the Strauss Commission (1996) laid the foundation for the reform of rural financial markets in South Africa. The recommendations impact more on national than on regional level. Neither did the commission deal sufficiently with the retail level to provide details of how to achieve increased access to rural financial services.

The micro-lending industry, servicing the needs of the '*unbanked*', show strong growth. Nearly 80% of adult South Africans have no access to conventional financial services and are regarded by traditional financing institutions as '*unbankable*' (Table 8.1). The major banks currently attempt to integrate these financial services into the broader, formal financial market while ensuring that market driven approaches are followed. The lack of rural infrastructure and high transaction costs might still leave large numbers of the rural poor without proper financial services. The trend amongst the major banks to rationalise and centralise certain services in rural areas as a cost saving measure might serve as an indication that they rather prefer not to serve the rural poor or alternatively, to serve them with other banking structures or institutions. According to Coetzee *et al.* (2003) '*not served*' has two broad explanations:

> First, not having access due to a lack of physical access, implying that the institution is too far away and that client access comes at tremendous levels of transaction costs, mostly borne by the client. Secondly, it may imply that the physical proximity of the institution is not the problem, but that the products and services offered and the delivery systems are not matched to client demand.

It is estimated that the informal small-scale financial market could be as big as six million people. About two million South Africans currently have micro loans (with an average loan size of R1,186), contributing to a R14 billion turnover. Approximately 2,000 micro-lenders are registered in South Africa. The introduction of a regulatory environment in 1999 provided respectability and legality to this industry (Bbenkele, 2003; Coetzee *et al.*, 2003). This was a necessary forerunner to the entrance of mainstream banks into the market. The industry has been consolidating itself, with big players buying out or establishing smaller concerns. Ten banks are now registered as micro-lenders, controlling about half of the industry's loan book.

Table 8.1. Categories of bankable and unbankable people in South Africa, 2003 (Coetzee et al., 2003).

Segment	Size	Urban	Rural	Banked	Un-banked
Employed/partially employed	5.2 m	67%	33%	48%	52%
Unemployed/unsupported	5.0 m	90%	10%	22%	78%
Township youth	3.0 m	100%	-	30%	70%
Supported family	2.4 m	5%	95%	33%	67%
Pension/grant	2.2 m	55%	45%	27%	73%
Totals		67%	31%	33%	67%
	17.8 m	12.3 m	5.5 m	5.9 m	11.9 m

Financial institutions focusing on rural people in developing countries rely heavily on principles such as joint surety and group membership. Member based financial institutions have been active in South Africa since the late 1800's. German missionaries already applied the concept of credit unions around 1880 in KwaZulu-Natal. Miracle *et al.* (1980) described the concept of collective action used in different variants of informal rotating savings and credit institutions throughout Africa as well as in South Africa. Different institutions, based on these principals operate within South Africa. Table 8.2 shows details of the different rural/informal financial institutions. The next section deals with the evolution of these institutions since 1994.

Table 8.2. Informal financial institutions in South Africa, 2003 (Coetzee et al., 2003).

Institutional format	Number	Members	Savings	Credit	Source
Stokvels	800,000	8,250,000	R400 m/month	?	NASASA
Burial societies	Est 250,000	8,000,000	R200 m/month	?	FinMark
Financial services co-operatives	62	80,000	R40 m	?	NDA
Saccos	28	13,000	R20 m	?	

8.6.1 Burial societies[8]

Burial societies are directed at unemployed and low income poor people in urban and rural areas. The clientele of the burial societies are vulnerable to exploitation because they are less educated and are regarded as the poorest of the poor. Burial societies generally perform important functions as networks of support to its clients and are highly respected amongst the poor communities, irrespective of some examples of exploitation.

Women bear much of the burden of the household and seek both economic and social support in burial societies. It is therefore common to find more women members than men, especially in the rural areas. However, in urban areas, migrant men may form and join a burial society to get support in their economic struggles.

Two categories of burial societies developed recently. Traditional burial societies differ from commercial ones in terms of structure and operation and their objective is not profit making. African entrepreneurs recognised the profit-making potential in combination with the social and cultural attractions of this system and launched some as profit-making insurance schemes. These burial societies draw members together regularly and offer support and solace during cultural rituals of burial and mourning. Members of commercial burial societies, however, do not have control over the finances of the society. The officials of these societies are appointed and not elected and they have full control over the management of the society. The financial affairs of such a society are mostly closed off to scrutiny of ordinary members. This is in stark contrast to the openness and active participation of the traditional burial societies.

Although the different burial societies operate differently, members join them for the same major reason, which is to be able to afford a decent funeral. Monthly payments to a burial society are therefore viewed not only as a contribution, but also as a saving option. Monthly premiums are affordable and flexible and transactions are done informally, as opposed to the formal procedures and high rigid payments required by formal banks. The support of other members in the society is also highly valued by members, although it is recognised that all the promises are not always fully honored.

Estimated membership of burial societies during 2002 in South Africa was 8 million members (FinMark, 2002). Quantification of the actual size of the movement is not easy due to the fact that many societies operate informally and are not registered. The national average size of a society is about 80 members. The membership numbers is testimony of the importance of the burial society movement in the life of poor people in South Africa. It might not be sufficient to ensure its survival in its present form, but the formal financing

[8] Most of the information are obtained from research published by the DGRV (2003) and a study completed by Coetzee *et al.* (2003).

institutions can use some principles entrenched in the movement to improve their own services targeted at the poor.

A new phenomenon is that a few of the profit-oriented burial societies have started to make loans for other purposes, such as small business.

8.6.2 Rotating savings and credit associations or stokvels[9]

Ardener, quoted by Schrieder and Cuevas, (1992) define the Rotating Savings and Credit Association (ROSCA) as 'an association formed upon a core of participants who agree to make regular contributions to a fund which is given in whole or in part to each contributor in rotation'. The principle of ROSCA is well-known all over the world, but it is mostly used and described within the context of developing countries. The most common name for the ROSCA in South Africa is 'stokvel'. The stokvel is more common in the urban than in the rural areas.

Stokvels are informal solidarity or self-help groups offering nearly all the financial services and some social and cultural needs required by the poor. This includes rotating credit and savings, funeral insurance, social occasions, etc. According to the National Stokvel Association of South Africa (NASASA) there are an estimated 800,000 stokvels, burial societies, and rotating savings and credit associations in South Africa comprising about 8.25 million adults accounting for about R400 million a month in savings.

Management of stokvels is voluntary and they are allowed to take deposits from members because they are exempted under the Banks Act under the 'common bond' exemption[10]. Most of the funds collected within stokvels are also distributed during group meetings.

The National Stokvel Association of South Africa (NASASA) was set up in 1988 as a lobby and umbrella group to represent the interests of stokvels. It is a non-profit organisation that provides education and public awareness programmes to the low income communities, encouraging them to save with the self-managed stokvels. It also develops management-training materials for members of stokvels. Stokvels can affiliate to NASASA at a subscription fee of R1,500 and R850 annually thereafter. As an affiliate, a stokvel has to agree to operate their business according to a code of conduct drawn up by NASASA. NASASA has never developed into a regulatory institution and does not have the capacity to regulate members.

Most significant for the formal banking sector is that NASASA negotiated with commercial banks and other financiers to use the accumulated savings of individual stokvels as a basis

[9] Most of the information are obtained from research published by the DGRV (2003) and a study completed by Schoeman, Coetzee and Willemse (2003).
[10] GN 2173 of 14 December 1994.

for granting loans. This could formalise the link between the formal and informal banking sector.

8.6.3 Village Banks[11]

The entry of the International Fund for Agricultural Development (IFAD) in South Africa earmarked the initiation of the Village Bank project. The first two phases of the project saw the establishment of three village banks in the North West Province from 1994 to 1996. A consultative group was formed existing of representatives from various institutions in the North West province to advise and mobilise support for this model.

The inability of private sector financial institutions to provide cost effective financial services in rural areas was the main reason for the village bank concept. High transaction costs and the outflow of hard earned savings without any benefits to local communities initiate the need for alternative banking systems.

The concept of the Village Bank was to create a financial institution that would decrease transaction costs of savings, increase the circulation of resources within the communities, and that could provide loans for re-investment in the communities where the savings are mobilised from. The Village Bank system was built on member ownership and control and relied heavily on the experience gained from stokvels, burial societies and other collective action institutions. This was seen as the means with which rural community could get access to a comprehensive range of financial services at more affordable transaction costs.

The initial project was completed in 1996 and the Financial Services Association (FSA) as a centralised support structure was established. The first Village Banks were legally registered as service co-operatives. The FSA obtained recognition for the Village Bank concept under a special exemption from the Banks Act during March 1998 from the Registrar of Banks. Due to the self-regulating requirements as contained therein the FSA had to assume the responsibility of regulating its member Village Banks. A regulating framework was developed in conjunction with the Registrar of Banks which led to the formal acknowledgement by the Registrar of Banks of the FSA as a self-regulating body for its member Village Banks.

Expansion of the village bank concept commenced during 1999 with the assistance of a grant of R7,000,000 from the Department of Social Welfare for the establishment of 70 Village Banks in 7 provinces. This grant enables the FSA to establish proper structures and capacity. 29 Communities were assisted to successfully establish their own financial services co-operatives during the funding period. Due to the lack of proper management

[11] Most of the information are obtained from research published by the DGRV (2003) and a study completed by Coetzee *et al.* (2003).

and inexperienced staff, funding was not renewed resulting in the operational closure of the FSA in 2001/2002.

Another village bank model called the FINASOL model developed from the South African Sugar Association's Financial Aid Fund (FAF). The original intension was to transform FAF funding into Village Banks registered as financial service co-operatives.

USAID supported the project and FINASOL was registered in January 1999 as a non-profit association under Section 21 of the South African Companies Act. The FINASOL model was based on a 'franchising system' whereby FINASOL was linked to Village Banks established on the same principles as developed in the Village Bank pilot project. The franchise system requires high levels of integrity, standardised procedures and policies as well as products from Village Banks. In return, these banks receive start-up capital to kick-start the Village Banks.

FINASOL was also recognised by the Registrar of Banks as a regulatory body with more or less the same functions previously allocated to the FSA. Due to a lack of funding the FINASOL operations seized to exist from July 2002. Most of the Village Banks established by FINASOL however continued to operate, as was the case with those established by FSA.

Currently no formal support services are rendered to the 62 Village Banks registered with the Registrar of Co-operatives. Thirty-two of these were members of FSA and 30 were members of FINASOL. The Co-operative Banks Act (Act 40 0f 2007) was signed by the President on 22 February 2008. This Act formalises and place new impetus on the number of applications to the Registrar of Co-operatives for registration as financial co-operatives. Many other rural communities are presently organising themselves with the purpose to register community based Village Banks. Membership of the existing Village Banks is estimated at between 60,000 to 80,000 members with a total portfolio of between R30 million and R40 million. Some Village Banks are continuing to expand their portfolios. One example is Motswedi with 1,552 members and a savings portfolio of R1.8 million.

The National Department of Agriculture (NDA) presently support the activities of the Village Banks. The grass root need for a Village Banking system is tremendous, so are the need to empower and train communities to manage these structures properly at community level.

8.6.4 Accumulating Saving and Credit Association

Accumulating Saving and Credit Associations (ASCAs) are essentially informal, small-scale, localised financial services associations. The main difference with stokvels is that ASCAs don't distribute regular payouts, but establish a pool of accumulated savings out of which loans are made to members on demand. ASCAs are less common than ROSCAs, partly because they require higher skill levels than the latter. Thousands of ASCA's are operating in South Africa. The Homeless People's Federation, for example, is based on ASCAs, and has over a thousand members in all parts of the country. Several newer NGO initiatives in the Western and Northern Capes have begun to support networks of ASCAs.

ASCAs' most significant features are that they are cheap to operate, they provide access to banking facilities on a collective basis and they build social assets, particularly amongst women because most ASCA's are founded and run by women.

8.6.5 Credit Unions and Savings and Credit Co-operatives

Savings and Credit Co-operatives (Sacco) are financial self-help and member driven organisations. This means that a Sacco sought its own funding from its members and re-apply it back into the community. The controlling body for Sacco's in South Africa is the South African Credit Co-operative League (SACCOL) that subscribe to recognised principles of the international Credit Union movement.

SACCOL was established in 1993 by the savings and credit co-operatives and credit unions in South Africa. SACCOL is owned and controlled by its member credit unions, which exercise proportional voting rights according to the size of membership. SACCOL is recognised as the representative of savings and credit co-operatives and credit unions in South Africa in Government Notice 2173 of 14 December 1994.

SACCOL received funding to the amount of US$ 1.3 million from 1994 until 2001 and adopted a self-sufficiency strategy. However, self-sufficiency levels were very low when USAID seized their support which compelled the movement to change its strategy to a bottom-up approach. Staff numbers at SACCOL level were reduced from 15 to only 4 which sees an increase in self-sufficiency from only 2% to 80%. Most of these staff was re-deployed by the individual Sacco's.

8.6.6 Summary of 'alternative' rural financial institutions

Table 8.3 is a comparison of 'alternative' institutional types and services provided in rural areas (Coetzee *et al.*, 2003).

8.7 Recommendations

8.7.1 Institutional setup and management concerns

The situation regarding the smaller financial services suppliers, including the informal entities involved in supplying financial services to small-scale operators in farming and related activities, has become rather disorganised. This is illustrated by the termination of operations by the Financial Services Association (FSA) and FINASOL. The supply of

Table 8.3. Comparison of 'alternative' institutional types and services provided (Coetzee et al.*, 2003).*

Service	Stokvels	Burial societies	Village banks	Saccos	NGOs
Saving	mostly rotation savings	mostly target savings to cover burial cost	unlinked savings	mostly credit linked savings	mostly forced savings
Credit	some inter-group lending	no credit unless also operating as a stokvel	subject to risk capital and leverage ability	subject to risk capital and leverage ability	savings linked credit
Transmissions	no transmissions	no transmissions	possible with link-bank relationship	possible with link-bank relationship	no transmissions
Cash transaction	no cash transactions	cash collections and payments	current account	no cash transactions	no cash transactions
Insurance	planned group scheme	mostly not underwritten	group or individual schemes	group scheme	no insurance
Economic empowerment	group and individual empowerment	group and individual empowerment	group and individual empowerment	group and individual empowerment	individual empowerment
Social empowerment	personal cohesion	personal cohesion	sustainable group structure	sustainable group structure	opportunity cohesion
Leverage	ability low	no leverage	ability high	ability high	no leverage
Market	unlimited	unlimited	primarily rural	primarily urban	mission target
Origin	organic	organic	external intervention	external intervention	external intervention
Gender	open	open	open	open	mission target
Legal entity	no legal entity	no legal entity	co-operative	co-operative	no legal entity
Ownership	informal	informal	formal shares	formal shares	informal
Group dependency	group specific	group specific	not a specific group	not a specific group	group specific
Group use of a formal bank	group account	group account	group account	group account	mostly no group account
Individual use of a formal bank	minority, not compulsory	minority, not compulsory	minority, not compulsory	majority, mostly compulsory	minority, not compulsory
Accommodation of subgroups	not	not	encouraged also include stokvels	not	not
Physical facility	informal	informal	formal	formal	informal
Cohesion	inter-group	inter-group	members to bank	members to bank	inter-group
Financial sustainability	all cost absorbed by the group	all cost absorbed by the group	income dependence	income dependence	cost absorbed by the NGO
Social sustainability	group ownership and control	group ownership and control	group ownership and control	group ownership and control	benefit dependence

financial services will probably improve in many respects, should some institutional network be formed to aid the operator(s) provided this network will effectively promote the interests of the participants, be well managed and deliver services – such as advice, training and representative services – to the participants. This network should be free from interference from governmental bodies and the commercial banking sector (which in South Africa is highly concentrated), although some linkages will be advantageous. There is a question as to the most appropriate form of organisation for lender groups whose members are the main source of funds.

Small to medium sized cooperatives may form a good institutional form, and such cooperatives may form a network; the network may establish a unit that can supply the needed services to the member cooperatives. In some European countries, such loan cooperatives have over time evolved into specialised credit institutions which cooperate mutually but retain their decentralised nature. This decentralised structure has been an effective management tool (Neveau, 1981). They have contributed to local saving and to local development (Huppi and Feder, 1990). One advantage of such a setup will be that a larger part of local savings will be utilised for the development of local entrepreneurs, including small-scale farmers.

8.7.2 Conditional credit to participants in marketing contracts

Experience has shown that in cases where firms supply credit to smallholders conditional to the sales of products to the firm (see Chapter 6), this system works well provided relations between the contracting firm on one hand and the smallholders on the other are and remain sound. Smallholder producers in such an arrangement should however also be made aware of the advisability of accumulating savings for purposes not covered by these contracts. This may involve other economic activity, household and family needs, and provisions for a possible future end of contracting arrangements.

8.7.3 The Land Bank

The Land Bank has over time mainly been involved with longer term loans, mainly to commercial farmers, largely for financing land purchases. Land thus bought has been used as collateral. This does by the nature of tenure relationships exclude communal land. The Land Bank has also over time indirectly supplied short-term credit to commercial farmers through loans to agricultural cooperatives. This scheme backfired, mainly in the 1980's and the 1990's, when indiscriminate loan extensions by some cooperatives caused farmers involved and some cooperatives themselves to default on loans. A repetition of such a system in the way it was operated can certainly not be recommended. Poor management also caused drastic deterioration in the condition of the LandBank in much of the first decade of the 21st century. Control over the Land Bank was transferred from the Department of Agriculture to the National Treasury, whereafter a turnabout strategy has been instituted. Financial reports indicate a successful turnabout (Land Bank, 2010).

The newly revitalised Land Bank can and should play an important role in the sector involved with the supply of credit to small-scale farmers, not by the large scale supply of short-term credit as was done in the late 20th century, but rather by offering banking services to the smaller suppliers of credit to these farmers – and to other small-scale rural entrepreneurs. Local suppliers like stokvels, lending groups, small local cooperatives, etc. can thus benefit. In view of the seasonal pattern of flows of funds in agriculture, the Land Bank can in such a way improve the provisioning of financial services to operators in many parts of the country; it must be borne in mind that seasonal production patterns differ among regions.

8.7.4 Savings mobilisation

In all efforts to improve the financial service supplies to small farmers, those involved should always keep it in mind that savings are as important as credit; the mistakes of the supply-led approach to credit must not be repeated.

8.7.5 Accessibility: the danger of centralisation

Efforts of centralisation in financial services to small-scale producers must be resisted. It should be remembered that, as pointed out earlier in this chapter, distance and its effects on physical access and eventually on transaction costs, can act as an important barrier to the ability to obtain the necessary financial services. The emphasis in this field should be on decentralisation. Economies of scale in such a type of activity are often imaginary rather than real. A review and analysis of various empirical studies in a large number of economic activities have shown that the difference between enterprises does not primarily lie in economies of scale; it lies in the economies of quality of management (Groenewald, 1991). Under poor management, large prosperous firms do become insolvent, as was demonstrated in the world-wide recession in the first decade of this century.

8.8 Conclusion

Marketing of agricultural products remains a field in which both poor, small scale farmers and small scale market operators have to cope with very formidable impediments. Access to appropriate finance is certainly one important impediment. As with other impediments, the problem of access to financial markets has to be solved by appropriate action and schemes by governments, but as important, by action within private sector operators and by small farmers and small-scale marketing operators themselves. Appropriate education and training is certainly prerequisite for such development.

References

Anonymous, 2008. Grameen Bank. Available at: www.wikipedia.org.

Adams, D.W., 1978. Mobilizing household savings through rural financial markets. Economic Development and Cultural Change 26: 547-560.

Adams, D.W. and R.C. Vogel, 1984. Rural financial markets in low-income countries. In: C.K. Eicker and M. Staatz (eds.) Agricultural development in the Third World. Johns Hopkins University Press, Baltimore, MD, USA.

Bahta, Y.T., 2003. Village banks, group credit, farmers' domestic savings mobilization in Eritrea. MSc (Agric) thesis, University of the Free State, Bloemfontein, South Africa.

Bbenkele, K.E., 2003. Rural finance expansion: experience in commercialization: selected South African micro finance case studies in rural financing. Micro Enterprise Alliance, Johannisburg, South Africa.

Bouman, F.J.A. and O. Hospes, 1994. Financial landscape reconstructed. In: F.J.A. Bouman O. Hospes (eds.) Financial landscapes reconstructed: the fine art of mapping development. Westview Press, Boulder, CO, USA, Chapter 1. Available at: http://edepot.wur.nl/134804.

Chandavarkar, A.R., 1988. The role of informal credit markets in support of micro business in developing countries. Paper at World Conference on Support for Micro Enterprises, Washington, DC, USA.

Coetzee, G.K., 1988. Die finansiering van kleinboer landbouproduksie in Suid-Afrika. MSc (Agric) thesis, University of Stellenbosch, Stellenbosch, South Africa.

Coetzee, G., J. Schoeman and R. Willemse, 2003. A review of the capacity, lessons learned and way forward for member based financial institutions in South Africa. Unpublished draft report, ECI Africa, Woodmead, South Africa.

Deutscher Genossenschafts- und RaffaisenVerband (DGRV), 2003. Burial societies in South Africa: history, function and scope. DGRV SA working paper series No 2. DGRV, Pretoria, South Africa. Available at: http://www.dgrvsa.co.za/publications.html.

Ellis, F., 1994. Agricultural policies in developing countries. Cambridge University Press, Cambridge, UK.

Fernando, N.A., 1991. Mobilizing rural savings in Papua New Guinea: myths, realities and needed policy reforms. Developing Economies 29: 44-53.

FinMark, 2002. The pro-poor microfinance sector in South Africa. Bay Research and Consultancy Services. Available at: http://www.finmark.org.za.

Fry, M.J., 1988. Money, interest and banking in economic development. The Johns Hopkins University Press, Baltimore, MD, USA.

Grameen Bank, 2008. Available at: http://www.grameen-ifo.org.

Groenewald, J.A., 1991. Returns to size and structure of agriculture: a suggested interpretation. Development Southern Africa 8: 329-323.

Heidhues, F., 1994. Consumption credit in rural financial market development. In: F.J.A. Bourman and O. Hospes, (eds.) Financial landscapes reconstructed: the fine art of mapping development. Westview Press, Boulder, MD, USA, Chapter 3. Available at: http://edepot.wur.nl/134804.

Hofmann, B., 1990. The rural household's financial requirement at formal and informal financial institutions analyzed in the Province Queme in the People's Republic of Benin. Standordgemasse in Westafrika – Zwischenbericht 1988-90. Sonderforschungsbericht 308. Universitat Hohenheim, Stuttgart, Germany.

Huppi, M. and G. Feder, 1990. The role of groups and credit cooperatives in rural lending. World Bank Research Observer 5: 187-204.

Kuhn, M.E., M.G. Darroch, G.F. Ortmann and D.G. Graham, 2000. Improving the provision of financial services to micro-entrepreneurs, emerging farmers and agribusiness: Lessons from KwaZulu-Natal. Agrekon 39: 68-81.

Land Bank, 2010. Annual report 2009/2010. Land Bank, Pretoria, South Africa.

Meyer, R.L., 1989. Mobilizing rural deposits: discovering the forgotten half of financial mediation. Development Southern Africa 6: 279-294.

Meyer, R.L. and G. Nagarajan, 1992. An assessment of the role of informal finance in the development process. In: G.H. Peters, B.F. Stanton and G.J. Fyles (eds.) Sustainable agricultural development: the role of international cooperation. Dartmouth Press, Aldershot, UK, pp. 644-654.

Miracle, M., D. Miracle and L. Cohen, 1980. Informal savings mobilization in Africa. Economic Development and Cultural Change 28: 701-724.

Neveau, A., 1981. Agricultural banks or multipurpose banks? In: International Association of Agricultural Economists (ed.) The rural challenge. Gower, London, UK.

Ong, M.L., D.W. Adams and I.J. Singh, 1976. voluntary rural savings in Taiwan, 1960-70. American Journal of Agricultural Economics 58: 579-582.

Padmanabhan, K.P., 1988. Rural credit: lessons for rural bankers and policy makers. Practical Action Publishers, Rugby, UK.

Schrieder, G. and C.E.C. Cuevas, 1992. Informal financial groups in Cameroon. In: D.W. Adams and D.H. Fitchett (eds.) Informal finance in low-income countries. Westview Press, Boulder, CO, USA, pp. 43-56.

Schrieder, G. and F. Heidhues, 1991. Finance and rural development in West Africa. Evaluation of financial projects for the rural poor with special emphasis on savings schemes. Proceedings of International Seminar in Quagadougou. University of Hohenheim, Stuttgart, Germany.

Spio, K., 1995. Rural household savings and consumption behaviour in South Africa. MSc (Agric) dissertation, University of Pretoria, Pretoria, South Africa.

Spio, K., 2003. The impact and accessibility of agricultural credit: a case study of small-scale farmers in the Northern Province of South Africa. PhD thesis, University of Pretoria, Pretoria, South Africa.

Spio, K. and J.A. Groenewald, 1998. Rural finance. In: J. Van Rooyen, J.A. Groenewald, S. Ngqangweni and T. Fenyes (eds.), Agricultural policy reform in South Africa. AIPA/Francolin Publishers, Cape Town, South Africa.

Spio, K., J.A. Groenewald and G.K. Coetzee, 1995. Savings mobilization in rural areas: lessons from experience. Agrekon 34: 254-259.

Strauss Commission, 1996. Interim report of the Commission of Enquiry into the provision of rural financial services (Strauss Commission). RP 38/1996. Government Printer, Pretoria, South Africa.

Vogel, R.C., 1984. Savings mobilization: the forgotten half. In: D. Adams (ed.) Undermining rural development with cheap credit. Westview Press, Boulder, CO, USA.

Von Pischke, J.D., 1991. Finance at the frontier: debt capacity and the role of credit in the private economy. The World Bank, EDI Development Studies, Washington, DC, USA.

Yunus, M., 2008. What is microcredit? Grameen Bank. Available at: http://www.grameen-info.org/bank.

9. Governance structures for supply chain management in the smallholder farming systems of South Africa

Ajuruchukwu Obi

9.1 Introduction and problem context

Of late, a lot has been said and written about governance in relation to the supply chains in which small producers are playing significant roles and necessitate adequate control and coordination to manage both domestic and donor resources to avert the persistent problem of insufficient market access (UN General Assembly, 2010; USAID, 2006). This has been largely in response to concerns about how power and control are exercised and the implications thereof for the efficient delivery of the produce of small farms and also how the welfare of both producers and consumers are affected. The more serious small farmers' problems have become in recent years, the more has the question of governance risen to the top of the agenda, and the reason is not difficult to find. Because of resource constraints and technical as well as institutional obstacles to production, the vulnerability of farmers has grown. The goal of the farm firm as a basic unit of economic activity is to maximise profits. But this goal is frustrated by the technical and institutional constraints confronting the farmers on a daily basis. On their part, the small farmers are adopting measures that minimise their vulnerability and enhance their welfare and profitability. Acting individually, each farmer adopts strategies that make him or her safer and more secure, regardless of what others are doing. The theory is that these actions can often make some people worse off while enhancing the livelihoods of others. This is clearly contradictory to the broader societal goals of promoting greater equity. This constitutes a market failure to the extent that less than optimal outcomes are realised by the actions that are designed to generate welfare enhancement for all. In fact, such market failures may often lead to considerable erosion of welfare for a large majority of the farmers or household units that are unable to compete in the unregulated competitive market (Obi, 2011).

Correcting the foregoing market failures constitute the major goal of public policy. The intention is to minimise the harm caused by the unbridled competition of disinterested producers and sellers. This is economic governance and involves the coordination of the actions of the numerous participants in the value chain to ensure that their interests are harmonised and they all work toward the collective good. But the way in which these are coordinated and channeled to achieve the goals of public policy can make all the difference between whether those goals are achieved or not. The belief is that the way transactions are managed and/or regulated affects the way and extent both producers and consumers benefit from the transactions, since the purpose of transactions is to enhance the welfare of all parties concerned.

Williamson (2009) has made a recent attempt to re-enact the work on economic organisation which Buchanan championed to argue that the fixation on optimisation and maximisation has downplayed the importance of 'mutuality of advantage' which is the fundamental reason why people engage in exchange. In his view, in which he was paraphrasing the views of earlier theorists, the primary unit of analysis is transactions, and our interest should be on how to coordinate the associated activities in such a way that initially divergent positions regarding relative value are reconciled so that all concerned are happy. It is this process of reconciling divergent positions in order to bring all parties to the point where they all benefit from the transaction that Williamson (2009) termed as governance.

Viewed from this standpoint, addressing the technical questions of farm production is not enough to get the farm and farm people to realise enhanced livelihoods and for the goal of poverty alleviation through improved market access to be realised; the way and manner that farm production and marketing are managed makes a world of difference. The crucial questions are who controls the chain and to what extent they do so? A related question would be what the implications of such controls are, in the sense of who gains how much and who loses how much from such an arrangement and in what ways the value/supply chains are affected by the nature of control and the sources of such controls. In the event the arrangements result in disadvantages or dangers to the performance of the chains, what remedial measures are available to deal with these and who exercises such remedies?

Understanding these issues is vital for the design of interventions to improve market access for smallholders because this forms the basis for drawing up the rule of engagement or guidelines for reconciling divergent positions in order to bring order to the market in the interest of all participants. The rules of engagement set for the industry provide the guidance needed to match results with expectations and require that all participants in any given chain observe them and submit to effective monitoring for the smooth functioning of the food system. Non-compliance with the rules of the game can often result, as Humphrey (2005) indicates, to inefficient distribution of goods and services and will create the sort of chaos that mirrors the perfectly competitive environment of classical and neoclassical economics. In such a situation, the goals of achieving enhanced welfare and associated gains are derailed.

The literature generally identifies three dimensions of value chain governance on the basis of the pattern of information flow within the chain, the rules and guidelines (both incipient and instituted) that prevail, and the mechanisms by which chain relationships are coordinated. In this sense, governance is a multi-faceted concept and does not necessarily imply a deliberate action to regulate the chain as it is often conceived. Rather, governance equally refers to the pre-existing situation in a particular chain in respect to the interactions between suppliers, consumers and other key players within the value structure. The concept recognises the market structure as it naturally exists and as it differs from one market type to another.

Value chain governance therefore mirrors the existing characteristics of the chain. The importance of value chain governance is emphasised by Gwynne (2006) whose work highlighted how financial, material and human resources are allocated within specific value chains. Gwynne (2006) also referred to a scenario whereby lead firms in an organisation exert authority on small firms as part of vertical coordination. Moreover, contracts are often used to tackle different aspects within the chain, involving quality, time of delivery and quantity among other aspects (Gwynne, 2006).

To understand the opportunities and limitations of smallholder integration in national or international food supply chains, a study was conducted to complement information available from the detailed studies reported in previous chapters. Previous studies have provided ample insight into the nature of local production of crops and livestock in this district of the Eastern Cape Province as well as the technical and institutional constraints faced by the smallholder producers (see Chapters 2 and 3 of this volume). Those studies addressed the micro-level dimensions of the food system which focused on, among other issues, the farm-level characteristics which govern or determine the level of physical production in the farming system and the obstacles encountered by producers in their efforts to meet the food and fibre needs of the population.

But sustaining those efforts has a lot to do with whether or not the participants in the system find them worthwhile. Invariably, this revolves around the question of incomes and livelihoods opportunities afforded by the farming enterprise. The extent to which smallholders participate in the food supply chains, that is, whether they are included or excluded from them, is therefore crucial to the debate about smallholder development and empowerment and helps our understanding of the role of markets and how they are organised to deliver enhanced livelihoods through promotion of access to markets for those who produce, as well as facilitating transactions that raise the real value of consumer incomes.

In order to fill this gap, a spatially broader but temporally limited enquiry was designed to identify the key determinants of the commodity mix in the farming system and the range of participants in the supply chain (that is producers as well as bulking agents and retailers), the existing regulatory environment, and the nature of the transactions that move foods along the supply chain between production and utilisation which were not examined in the previous studies.

9.2 Methodology and focus areas

Whereas the previous study for the Eastern Cape that provided the material for Chapters 3 and 4 focused solely on Nkonkobe Municipality, the need for a more policy-relevant enquiry into the supply chain to identify the existing governance structures meant that a wider geographical space should be defined. For this reason, the entire Amathole District was taken as the sample frame from which four municipalities, namely Amahlathi, Buffalo City, Mbashe and Nkonkobe local municipalities, were enumerated. Figures 9.1 present a map of the district showing the relevant local municipalities.

As already indicated, the SANPAD-funded research on 'Institutional and Technical Constraints to Smallholder Development', carried out a comprehensive inventory of the key components of the farming system, focusing specifically on the Nkonkobe municipality, and identified the institutional and technical constraints to smallholder participation in the food system. In order to understand how the smallholder producers are linked to supply chains and the processes that move the food along these supply chains, it was necessary to extend the dataset to incorporate details on what happens to the food between the farm and table and how that process is facilitated. In this regard, the inventory went on to identify those who sell what and to whom and how the foods are procured and the arrangements for their distribution within and beyond the local area, municipality, and the wider district. This exercise was conducted in the format of a comprehensive auditing of the structures and how they operate without necessarily looking at what their impacts are on the rest of the system. That aspect is taken up in Chapter 10 which examines the implications of the governance arrangements for market access. The primary information gathered for this chapter was complemented by further document analyses on agro supply chains, network issues, and the food markets in developing and emerging economies. The focus areas for the information gathering were carefully chosen to promote understanding about the needs

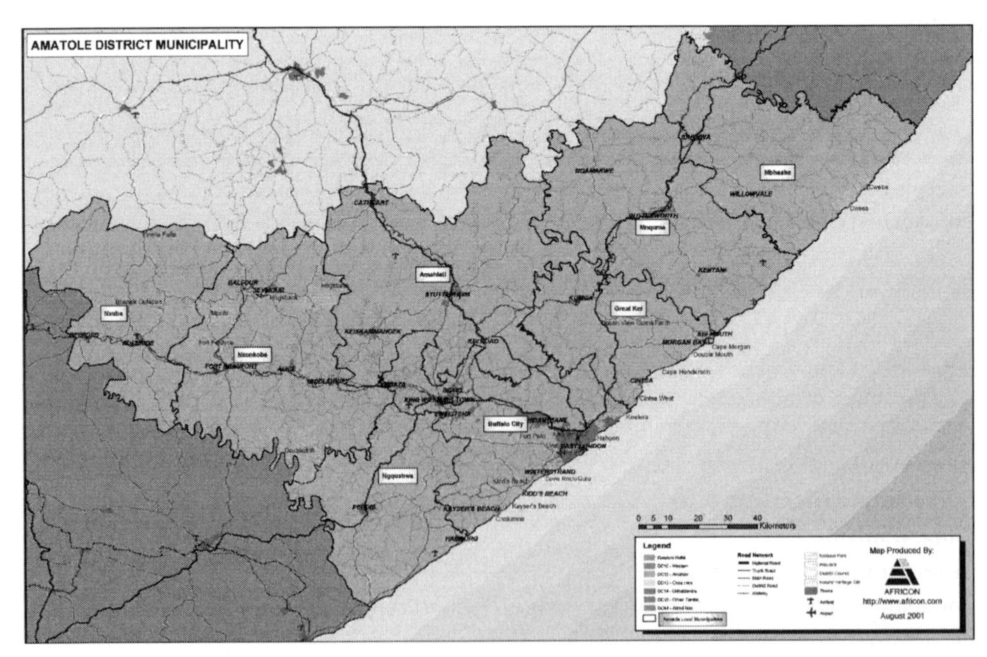

Figure 9.1. Geographical location of the Amathole District of the Eastern Cape Province, South Africa.

and aspirations of the farming households enumerated, what resources are available, the roles played by different participants in the chain, and how the benefits of the economic activities are distributed.

The information gathering exercise adopted a case study approach. The previous study identified the principal commodity categories of the district and these formed the basis for developing the survey instruments for the next step of enumerating individual production and consumption units as well as retail outlets. In this regard, a semi-structured questionnaire was drawn up and the following specific questions were included to afford deeper insights into the governance mechanisms that are important in the local food system of the district:

- What are the needs and aspirations of the farmers and other chain participants in the district?
- Sources and destinations of principal foods in the district – from where do the food retailers buy produce for sale in Alice/Fort Beaufort/Stutterheim/King Williams Town/Dutywa/Willowvale?
- Under what arrangements do the transactions take place? Do they buy from the small-scale producers in the surrounding villages or are these producers discriminated against on grounds of poor quality or insufficiency of output?
- Formal and informal producer-retailer contracts – what are the elements of the contract and how are they enforced, if they exist in the first place?
- Municipal regulatory mechanisms – trade licensing requirements – what state institutions and departments have responsibility for this, what are their requirements, how do they operate, etc.?
- Membership of marketing associations for eligibility to trade specified commodities? – who has access to the fresh produce market?
- What are the links between major retail chains and Fresh Foods Markets in the major market centres such as between the Fruit and Vegetable City in King Williams Town and East London? (Fruit and Vegetable City is a well known retail seller of fruit and vegetables in South Africa).
- For the small-scale citrus producers, the influence of the producer organisations – Citrus Growers Association – and the rules of bodies such as EurepGap and local is a Good Agricultural Practice (GAP) scheme aimed at ensuring food safety, environmental protection, occupational health, safety and welfare. Are these applicable in the district and in what ways are they applied?

In 2008 and 2009, during a one month period in each year, a data collecting procedure was instituted to enumerate the major food retail outlets in Alice, Fort Beaufort, Stutterheim, King Williams Town, Dutywa and in Willowvale in order to track their main suppliers in relation to the pre-identified principal commodity categories and build up deeper understanding of the rhythm and flow pattern of supply and sales, and the governance mechanisms influencing transactions in the food chain.

Of particular interest were such issues as supplier-retailer relationships and retailer-buyer/consumer relationships. The factors affecting these relationships were tracked through the semi-structured, multiple-visit interviews of the principal retail outlets, representative producers and representative consumers. For retailers, the interviews focused on identification of major suppliers, when and how often each commodity moves through the chain to capture the element of market timing, quality concerns if any, and whether or not the chain delivers the commodity when it is needed or whether some waiting period is involved. For producers, the questions included where they sold, how they transported the commodities between farm and shop, what motivates choice among alternative transport modes, and destination of produce, and whether or not there are well-defined arrangements and procedures for facilitating exchange. In the case of consumers, the questions revolved around where commodities are procured, at what prices the commodities are procured and how these compared with those prevailing elsewhere. It was important to determine whether or not consumers adjudged a particular outlet positively or negatively.

Figure 9.1 delineates the geographical boundaries of the Amathole District of the Eastern Cape Province. The main towns of Balfour, Seymour, Alice, Fort Beaufort, Healdtown, Middledrift, King Williams Town, Stutterheim, Dutywa, Willowvale, and East London fall within the district. Out of these, Alice/Fort Beaufort, Stutterheim, King Williams Town, Dutywa and Willowvale were closely enumerated while fleeting, rapid assessments were conducted in the rest of the towns. The dominant geographic characteristics of this area include the extensive mountain range of the Amathole which encloses a valley of about 80 km length being part of a vast catchment area stretching about 1,715 square kilometers across much of the southern parts of the Eastern Cape Province. The district retains an enormous amount of primary forest, especially in the Hogsback and Keiskammahoek areas and is drained by a large number of rivers, notably the Keiskamma, Buffalo, Nahoon, and Gqunube which flow in a markedly south-easterly direction to empty into the Indian Ocean. Much of the area is endowed with fertile soils which support intense agricultural activity. A sizeable population of about 1.7 million in mid-2007 across the 8 municipalities of the district (Table 9.1) suggests immense market potential for agriculture.

9.3 Needs and aspirations of chain participants

In reviewing the literature on governance, including work done by Professor Nawar in Egypt, Mdaka (2010) observed that one of the key elements of governance is decision-making on how to go about tackling the problems that individuals and groups face from day to day. This implies that in economic production, the human dimension is crucial because that is what determines what activities take place in the first instance. Whether or not local people produce for the market or for home consumption or even produce at all, is determined by the prevalent value system. What people consider normative for any particular environment is a function of their value system and this should therefore be the starting point for the consideration of existing governance structures in any particular

Table 9.1. Population distribution of the Amathole district of Eastern Cape Province, 2007 (Amathole Economic Development Agency, Annual Report, 2006/2007).

Municipality	Population figure
Amahlathi	139,000
Buffalo City	703,000
Great Kei	45,000
Mbashe	254,000
Mnquma	288,000
Ngqushwa	84,200
Nkonkobe	123,000
Nxuba	25,000
Total	1,665,000

environment. Bembridge (1986) has similarly observed that the human element is a key factor in agricultural development although farm management research and policy do not accord it the necessary recognition. Without question, land, technology and capital are important. But utilising these resources efficiently requires that they be developed, organised and operated. Optimum productivity depends on rational decision making, which, in turn, depends on the personal and socio-psychological characteristics of the farming population.

One way to operationalise the human element in local economic participation is to assess the needs and aspirations of the farmers and household members. The needs of the farmers are expressed through their production goals while their aspirations define the options they are willing to consider to address those needs. The farming goals reflect the production possibilities on the basis of available resources and know-how in the current production period. On the other hand, the farmer's aspiration refers to what the farmer considers as the options that can be pursued to improve any perceived undesirable current situation. Accordingly, the information gathering exercise interrogated the farmers and households about these issues. Two categories of smallholder farming can be delineated in the district, namely the relatively large holdings devoted to tree-crop farming, mainly citrus, in the former Ciskei area, and the smallholder arable crop farming, including home gardening, across the district. This aspect of the study focuses on the latter. It was found that farmers' goals and aspirations differed considerably and reflected different degrees of agricultural commercialisation. For instance, the indication is that goals ranged from 'not farming' to 'production for home consumption', production for 'marketing', to a combination of both marketing and consumption. On the other, aspirations ranged from an unwillingness to farm, an unwillingness to increase production, and an eagerness to increase production. The results of this investigation are summarised in Tables 9.2 and 9.3.

Table 9.2. Summary of farmers' goals (adapted from Muchara et al.*, 2010).*

Status	Number of respondents	% of total
Not farming	2	2.4
Marketing	0	0
Consumption	74	90.2
Marketing & consumption	6	7.3
Total	82	100

Table 9.3. Summary of farmers' aspirations (adapted from Muchara et al.*, 2010).*

Status	Number of respondents	% of total
Not farming	2	3.0
Not willing to increase production	34	41.0
Eager to increase production	46	56.0
Total	82	100.0

The indication from Tables 9.2 and 9.3 is that much of crop production is oriented towards home consumption rather than to the market. According to the results, as many as 90% of the farmers enumerated produce solely for home consumption. But the majority of the farmers recognise the importance of increased production as the means to improve their situation. For instance, 56% of the farming households expressed an eagerness to expand production while about 42% were satisfied with their existing production levels. An insignificantly small proportion of the sample did not see any need to farm at all. These were probably the same households that did not farm at all and never defined their goals in terms of farming. Most villagers revealed that their current production levels were inadequate to last them to the next season, hence a need to produce more for food security reasons. In further assessing the human element in resource control and management and economic participation, the role of group membership in the pattern of needs and aspiration was investigated and the results are presented in Table 9.4.

According to the results, the majority of the farmers who produced solely for home consumption did not belong to a project or farmer association. There was no evidence of full commercialisation within the sample, but a few farmers who belonged to a project combined marketing with home consumption.

Table 9.4. Influence of membership of irrigation projects on household goals in crop production (adapted from Muchara et al.*, 2010).*

Status	Goals for crop production				Total
	No crops	Consumption only	Marketing + consumption	Marketing only	
Non-members	1	43	0	0	44
Members	1	31	6	0	38
Total	2	74	6	0	82

Small-scale livestock production was also investigated and the results are presented in Tables 9.5 and 9.6. Again, important differences were observed between households in terms of goal definition and aspirations. For instance, it was shown that livestock production goals ranged from 'no livestock', 'cultural purposes', 'consumption', 'marketing and consumption', and 'marketing'. The indication is that most households keep livestock for home consumption

Table 9.5. Farmers' goals in relation to livestock production (adapted from Muchara et al.*, 2010).*

Status	Number of respondents	% of total
No livestock	33	40.0
Cultural purposes	3	4.0
Consumption	36	44.0
Marketing & consumption	10	12.0
Marketing	0	0
Total	82	100

Table 9.6. Farmers' aspirations in relation to livestock production (adapted from Muchara et al.*, 2010).*

Status	Number of Respondents	% of Total
No livestock	33	40
Not aspiring to increase production	8	10
Aspiring to increase production	41	50
Total	82	100

and only a few engage is some livestock sales. But a good number of the respondents were willing to expand their herd probably to increase their protein consumption. Still a sizeable number did not see livestock production as a viable option for enhancing livelihood.

9.4 Resources and opportunities in the food supply chains

The other important element of governance is the resource profile of the key chain participants. The results of the analysis on opportunities and limitations of smallholder integration in food supply chains are summarised along with discussion on their implications for enhanced livelihoods by contributing to poverty alleviation. The next section looks specifically at the identified commodity mixes. Then the identified chain participants are presented and discussed in relation to their scale of operation, enterprise specialisation, needs and aspirations, and control over the chain. Following this, the chapter examines the different types of marketing channels utilised by the smallholders in the district. Two main types of farming organisation were specifically enumerated to isolate the role of alternative coordination and control mechanisms on the farming systems and the implications for different categories of chain participants. At the one end are the small-scale producers cultivating small-sized individual plots and herding a small number of livestock, principally small-stock. This form of production organisation was largely oriented towards subsistence production with little or no surpluses generated for markets. At the other end of the spectrum are the relatively larger holdings which may either be organised as individual units or as group operations, generally around government-funded agricultural projects commonly linked to irrigation schemes. Naturally, this allows for the prevailing governance mechanisms to be identified and assessed for their implications for smallholder participation in domestic and international supply chains.

9.4.1 Principal food products and commodity mixes

The food system features a broad range of food types ranging from vegetables, pulses, grains such as maize and the legumes (including sugar beans, peas, etc.), citrus and deciduous fruits, as well as livestock, meat and other livestock products.

Crops

Despite the dominance of relatively small-scale individual operations in what is now typically described as 'keyhole gardens' or 'homestead gardens', the system features a certain vibrancy that mirrors the diverse racial and ethnic composition of the district with its attendant diversity of tastes and preferences. The small homestead gardens that produce at subsistence levels were mainly producing vegetables. The relatively larger individual small holdings were also concentrating on vegetable production but were also involved in the fruit tree sub-sector, notably citrus. Overall, for the former Ciskei area, the main food products identified include fruits and vegetables such as cabbages, bananas, peaches, apples, citrus products, tomatoes, onions, butternut, peas, beetroot, potatoes, spinach and other leafy vegetables, watermelon, pumpkins, etc. In the part of the district falling under the former Transkei, it was observed that the scale and intensity of arable farming were different and the orientation of production towards the market seemed more noticeable than in the case of the former Ciskei homeland area. Similarly, the tendency to engage in group farming seemed more common in the former Transkei area than the Ciskei area. As a result, group farming projects in the former Transkei area were investigated to identify any distinctive features in their operations that contribute to the explanation of differential access to profitable supply chains. Two such groups selected were the Foundation Community Project and the Ciko Community Project, both in the Mbashe local municipality. The two projects make use of sprinkler irrigation to produce a wide range of crops, including cabbage, spinach, butternut, broccoli, pumpkins, potatoes, green paper, carrots and maize, onions, and tomatoes. On the basis of the foregoing, about 7 agro-food commodity chains can be distinguished for the dominant food crops produced in the Amathole District. These are illustrated in Table 9.7.

Among the food grains, it was found that the most popular items include maize, sugar beans, peas, etc. An interesting feature of the system is the diversity of forms in which foods are prepared and consumed.

Livestock

Livestock was found to occupy a special place in the agricultural and food system of the Nkonkobe municipality as well as the larger Amathole district. The households keep a wide variety of livestock including cattle, sheep, goats, pigs, and poultry. Reasons for keeping livestock ranged from their use for draught power, to household consumption as food, live sales to raise cash, traditional ceremonies such as payment of bride price (*lobola*) and rituals, notably to offer sacrifices to ancestors for various reasons. A variety of livestock products were also identified, including beef, pork, mutton, bacon, chicken, eggs, dairy products, wool and mohair, etc. which are either consumed by the household or sold for cash to meet household needs. The livestock chains identified in the study area are presented in Table 9.8.

An interesting feature of the livestock production system in the communal areas is the fact that whereas ownership of livestock is organised at the household level, their rearing/grazing is organised communally on joint property over which the individual household has no control. Conflicts do arise in this situation because private goals to optimise returns from livestock enterprise often result in the indiscriminate exploitation of the communal resource. Local associations regulating activities of livestock owners operate but problems remain. Farmers/households also complain of the rampant theft of small stock which is adversely affecting the popularity of small stock ownership in the area. This latter problem is so severe in a place like Keiskammahoek of the Amathole district municipality that households have largely abandoned the idea of keeping sheep and goats because their small size makes them a convenient target for stock theft.

Table 9.7. Selected crop food chains in Amathole district of the Eastern Cape Province, 2008-2010 (Field surveys 2008 and 2010).

Chain	Commodity
Citrus chain	oranges
	naartjies
	grapefruit
	fruit juice
Potatoes chain	potatoes
	frozen potatoes
	potato chips
Tomatoes chain	tomatoes
	tomato juice
	tomato paste
	peeled tomato
Maize chain	maize
	maize flour
	roasted maize
	popcorn
	boiled maize
Vegetable chain	spinach and other leafy vegetables
	onions
	watermelon
	cabbages
	pumpkins
Fruit chain	apples
	peaches
Grain chain	sugar beans
	peas

9.4.2 Chain participants

A striking feature of the local food system of the Amathole District of the Eastern Cape Province is the large number of operators involved in the supply chain. At the primary production level of the chain are the small farmers and home gardeners who have been described in the previous sections and are spread all over the communal areas of the District. In addition to this category, there are numerous projects funded under assorted government and non-governmental programmes that involve smallholder farmers as beneficiaries. These project farming schemes are being implemented under various contractual arrangements

Table 9.8. Selected livestock chains in Amathole district, 2008-2010 (Field surveys, 2008 and 2010).

Chain	Product
Beef chain	cattle
	beef and veal
	biltong (dried and smoked beef)
	braai (barbecue)
	milk and dairy products
Pork chain	pigs
	pork chops
	bacon
Mutton chain	sheep and goats
	mutton
	lamb
	wool and mohair
Poultry chain	live chicken/turkey
	frozen chicken/turkey
	eggs

that combine technical support at the production level and advisory services and assistance with marketing of the farm produce. Among such schemes, the Zanyokwe Irrigation Scheme stands out and will be described in some detail. At the other end of the chain are the food retailers who are linked to the producers through 'middlemen' and other genres of bulking agents and intermediaries. The study also identified some farmers' associations operating in the area and providing one form of support or the other to the smallholder farmers in their production and marketing activities.

On the basis of discussions with chain participants and others with a relevant knowledge, it is clear that distinct links exist between producers and suppliers and the final users of the smallholder produce. In addition, discussions revealed that there are interactions with formal and informal regulatory mechanisms and intermediation structures, including marketing/ producer associations which ensure the smooth-running of the system.

Primary producers

As indicated above, the Nkonkobe food system features a large number of producers who operate small farms within the communal areas of the municipality and employ largely traditional technologies to grow a diverse range of crops. Past studies have characterised the smallholder sector as being dominated by mainly black households, producing on relatively small plots of land, with limited resources for household subsistence or sale (Hebinck and Lent, 2007; Kirsten and Moldenhauer, 2006; Oettle *et al.*, 1998). There has been a considerable amount of interest in this category of farmers in recent years, with many projects being set up to address their unique problems of low production and productivity which is often blamed on inadequate market access. In collaboration with the Water Research Commission, researchers at the Faculty of Science and Agriculture of the University of Fort Hare have been implementing a project to introduce water harvesting technologies to the farmers to address the problem of acute water shortage for both farming and domestic use. At the other end of small-scale production are the relatively larger operators who cultivate larger holdings, keep more animals, and whose production is more oriented to the market and is thus better linked with the local and international supply chains (Chikazunga *et al.*, 2007). These different categories will be described in turn.

Small-scale gardeners

An important element in the participation of the primary producers in the supply chain is their scale of operation which has implications for the level of output they are able to deliver. At one end of the continuum are those whose operations are so small that they can only be described as 'gardeners' with small farm plots located around the homesteads and managed entirely by the farmer. These plots have recently been tagged 'keyhole gardens' (Mokitimi, 2007). On such plots, the gardener typically plants a range of vegetables such as cabbages, lettuce, onions, spinach, etc. As might be expected, both quality and quantity are inconsistent at this level and the arrangements for marketing are equally not formalised. Sometimes surplus produce is sold at the homestead, usually to itinerant bulking agents who visit the homestead to enquire as to the availability of such surpluses based on prior relationships. At other times, the producers carry their meagre produce to the village spazas (small convenience shops located in the villages). Infrequently, the meagre surpluses are conveyed by taxi to the town of Alice where they are sold directly to consumers at makeshift stalls set up on street corners and at the taxi rank just below the bridge linking the town of Alice and the University of Fort Hare.

A study (Mzamo *et al.*, 2007) in six villages of Alice, namely Nkobonkobo, Mgquba, Lenge, MaBheleni, Esikolweni, and Mavuso, revealed that the majority of the households fell within the above category of homestead gardeners who grew assorted vegetables, legumes and food grains, but also kept a small number of livestock. The study found that only about 5% of the households kept more than 10 cattle while the majority kept between one and 10 cattle. Most households actually kept less than 4 cattle. The majority of households that owned livestock kept small animals, mostly sheep, goats, poultry and pigs, with some households keeping as many as 30-36 animals. In another study conducted in two communities (Cildara and Kubusi), both within the Amathole District of the Eastern Cape Province, it was found that the local people engage in livestock production for multiple reasons (Ngxetwane, 2008). In these cases, many of the livestock producers owned about 7 goats, 9 sheep, 10 cattle, although wide variations were also observed (Ngxetwane, 2008). A pointer to the small size of their operations is the level of gross margins realised by each household. For instance, the study determined that gross annual earnings from farming averaged about R609.12 (or US$ 63) for male-headed households and R366 (or US$ 38) for female-headed households, reflecting their differential access to key productive resources such as land, among other handicaps. In respect to this particular asset, there seemed to be important gender-related differences in the possession of title deeds to farmland which has implications for access to institutional financial resources.

Among the small-scale farmers are those who are participating in agricultural development projects being implemented under diverse governmental and non-governmental schemes. The Zanyokwe Irrigation Scheme deserves special mention because of its status as a virtual fore-runner in the black empowerment schemes even at the height of the Apartheid policies. In one review of this scheme, WRC (2005) provides insights into the background to its formation by the erstwhile Ciskei Homeland Government in 1984 with the goal of improving rural livelihoods in that part of South Africa at the height of the Apartheid policies. According to the assessment, the scheme was designed for 80 small scale farmers distributed across 6 villages, Zinguka, Ngqumeya, Zanyokwe, Lenye, Burnshill, and Cwaru. A study conducted in 2007 by Mnduze and Njajula (2007) revealed that the farmers were devoting a decreasing proportion of their land-holding to the contract farming arrangement. There is a feeling that this trend is as a result of limited opportunities to market the larger output from contract farming. In terms of the demographic character of the contract farmers, it was shown that the majority of them are male, middle-aged and with about 7 years of formal schooling. It was also shown that plot sizes are quite small on average and the farmers continue to face the constraints of not having title deeds to the lands they cultivate.

Emerging farmers

Other producers operating in the chain at a relatively higher scale than the 'keyhole gardeners' include the small-scale crop and livestock producers in Alice town and the few black small-scale commercial farmers officially referred to as 'emerging farmers'. Within the Nkonkobe municipality, the most prominent of these emerging farmers are those growing citrus on much larger holdings in the Balfour and Seymour areas of the municipality. According to Mather (2004), citrus farms range in size from 0.5 ha to 500 ha within the small-scale sector, whereas the large farms can go up to 6,000 ha. These citrus farmers also grow some field crops and keep a few animals in many cases and are characterised by the larger size of their operations and their greater market-orientation. The participation of this category of producers in the food supply chains has interesting features that are discussed in some detail.

The citrus producers in the Kat River Valley afford more interesting characteristics for analysis of supply chain participation in the Nkonkobe Municipality because the area hosts both small-scale and large-scale producers and includes the range of activities involved in citrus production and marketing. For this reason, the citrus producers of the area are exposed to the kind of diversity of conditions not easily found elsewhere and within easy proximity of the University where the researchers are based.

In an effort to integrate the black population into South Africa's commercial agriculture, the former Homeland Government of the Ciskei in 1988 established 22 black farmers on about 460 ha of land appropriated from former white owners for purposes of citrus production (Umhlaba Consulting Group, 2006). While it was originally intended that these farmers would have full ownership of the land within 5 years, this has not happened and the farmers still do not have titles to the land in the two decades since the initial transfer. As might be expected, the lack of title deeds complicates the farmers' current difficulties in accessing institutional credit.

The credit constraint faced by the emerging citrus producers has been manifested largely in the depression of their output which invariably falls below the critical minimum in terms of both quantity and quality required to capture secure markets. These farmers report that they are unable to timeously procure needed inputs or carry out critical farm operations. Even when output is realised at levels that are sufficient for market delivery, the emerging citrus farmers confront the problem of inadequate transport facilities; badly maintained vehicles plying on poorly maintained roads which affect the quality of output delivered to the markets.

9.5 Distribution and retail outlets

Another key element of governance is the distribution of benefits. As might be expected, the distribution of benefits will differ with the nature of products, the characteristics of the producers and the size of markets which has implications for the incentive structures and whether or not producers consider it worth their while to participate in a particular production activity. With this in mind, an inventory of the distribution and retail outlets in the smallholder sector of the district was undertaken. The major areas covered by the assessment were Alice, Fort Beaufort, and King Williams Town, Willowvale and Dutywa.

The major fresh produce retailers in Alice include the Spar supermarket operating in the Kwantu Shopping Complex, the Royal Supermarket, Kwamalinga Butchery, Pack Food Supermarket, Fresh Farm Chicken, Tremeer Butcher, Fruit & Veg Market, and Broadway Supermarket. In Fort Beaufort, the most notable retailers are Fruit and Vegetable City, Georgie Spar, Shoprite, Fruit and Vegetable Market, and numerous butcheries and vegetable shops. In Dutywa, the Georges Fruits and Vegetables shop is an important outlet used by the producers and consumers in that area. Several other shops including Spar, Boxer Super Stores, Food Town, Kwamadyasi, Emsengeni Wholesalers, Shoprite and Spargs supermarket which operate in the Mbashe local municipality were also identified and enumerated. In addition to these, there are numerous vendors, including those that operate on street corners in kiosks and open spaces. There are also the itinerant vendors who sell fresh produce and semi-processed foods from the trunks of motor vehicles, especially during the lunch hour and on weekends.

The relative importance of the different retail outlets in terms of their popularity among smallholder farmers has been studied extensively in South Africa. According to Chikazunga *et al.* (2007), smallholders make use of alternative retail outlets for the disposal of their farm produce. For the tomato sub-sector, a popular smallholder crop, it was found that in Limpopo and Mpumalanga provinces, there was a preference for the less formal retail outlets such as hawkers (Chikazunga *et al.* 2007). For instance, as much as 80% of the smallholders enumerated indicated that they sold their fresh produce to hawkers and slightly more than 10% sold to the formal markets such as agro-processors, supermarkets, etc. (Chikazunga *et al.* 2007). Table 9.9 shows the distribution of two categories of smallholder tomato producers (traditional and modern) by their preferred marketing channels in two provinces in 2007.

According to Table 9.9, the less formal channels such as hawkers and the local open-air markets were more popular among both the traditional and modern smallholders than the formal channels such as supermarkets, agro-processing firms, and the National Fresh Produce Markets (NFPMs). But the results showed that the modern smallholders, that is those who use more advanced farming techniques and are relatively better capitalised and resourced than the rest, were marginally more keen to use the formal channels. The reasons for specific channel choices have been explored elsewhere (Chikazunga *et al.* 2007), and often include the fact that hawkers offer better prices than the formal outlets such as agro-processors and supermarkets who also tend to demand more stringent quality standards than the smallholders can afford.

Table 9.9. Distribution of smallholder tomato producers by preferred marketing channels in Limpopo and Mpumalanga Provinces, South Africa, 2007 (Adapted from Chikazunga et al.*, 2007).*

Marketing channel	Traditional producers (%)	Modern producers (%)
Supermarket	1.0	3.0
Processors	2.0	6.0
Fresh produce market	3.0	9.0
Hawkers-on-foot	36.0	30.0
Hawkers-on-wheels	34.0	29.0
Local open-air markets	24.0	23.0
Total	100.0	100.0

The sources of the produce sold in these Amathole District markets differed considerably. The larger retail outlets and supermarkets obtained their stocks from large bulking centres in East London such as the Buffalo City Municipality Fresh Produce Market in East London. The produce are normally trucked to the shops in Alice and Fort Beaufort once or twice a week depending on the season, with the deliveries being more frequent during festive periods such as Christmas, Easter Holidays, etc. The smaller shops and retail outlets received their supplies from nearby King William's Town, especially from Mr. Potato (a fresh produce market), and from the Alice Market.

The special circumstance of predominantly low-income communal areas is illustrated by the experience in Mbozi and Ciko villages in the catchment areas of the Willowvale and Dutywa towns. These towns have fruit and vegetable (Fruit & Veg) shops whose main role is distribution of assorted agriculture commodities. The irrigation schemes in those villages produce vegetables but consumers and small shops in the area prefer to procure from the larger fresh produce market. For the small shops that might otherwise be expected to procure from the irrigation schemes, it was revealed that they only buy from those scheme members who are able to transport their commodities to the shop, thus highlighting the importance of resource profiles and the structure of incentives in governance of smallholder supply chains. Delivery is a real challenge with both Ciko and Foundation Community projects, who do not own any trailer or truck to transport their produce to the market. Limited financial returns and poor roads also prevent the projects from hiring transport to deliver their produce to the market.

The smaller shops serving low income communal areas are also in a special category and often have to make special arrangements to ensure participation in the supply chain. An interesting scenario was observed where some shops collectively hire trucks to collect fresh produce from the East London Municipality Market, but same shops are not prepared to hire transport to collect fresh produce from nearby (Foundation Project – 17 km; Ciko Project – 7 km) irrigation schemes. This is an indication of special requirements that these shops expect in order to reliably procure from a specific agricultural supplier. An interview with Mr. Marios (Box 9.1), who owns Ndubs fruits and vegetable shop at Willowvale, highlighted some aspects that need to be considered in building a strong base for smallholder agricultural projects.

The interview in Box 9.1 gives some insight into the importance of accessibility and how the infrastructure and resource profile affects smallholder farmers' participation in the supply chain. Fruit and vegetable shops can offer a good market for smallholder farmers especially in rural communities, where direct competition from large commercial farmers is assumed to be low unlike at the fresh produce markets. However farmers fail to utilise this opportunity due to resource constraints as well as production and market related factors.

Box 9.1. Interview with a fruit and vegetable shop owner (Muchara *et al.*, 2010).

Mr. Marios owns two shops at Butterworth and Willowvale town. Both shops market agricultural commodities (fresh produce). He does his procurement from East London Municipal Market. He does not own a truck. He combines his orders with three other shop owners who own a seven tonnes truck and a trailer. They collect fresh produce twice a week from the East London Market and pay R300 per trip (i.e. R600/week) for transport. Mr. Marios believes that the transport cost is very affordable considering the distance and the volume of produce they get per trip. They prefer the East London Market because their order is arranged by agencies before their collection date to avoid paying for empty trips. This is not possible with farmers who at times cannot fill a one tonne *bakkie* (pick-up truck) with produce and often demand a higher price than the fresh produce market for their products. Procuring from farmers is stressful and involves extra costs because one would need to employ more workers to do packaging of the products, buy own packaging material and high vehicle maintenance costs due to poor roads. He would rather travel 200 km by tar and collect bulk products than travel 10 km to get two bags of cabbages and lose his truck through breakdowns. He believes that he has a good competitive advantage over other dealers due to lower transport costs.

Production levels and product quality are some of the critical factors affecting relationship between smallholder irrigation farmers and the major buyers like fruit and vegetable shops. In 2008, the Foundation Community Project managed to supply their cabbages to Georges Fruit and Vegetable Shops at Dutywa and Butterworth. But there seems to be a thawing of the trading relationship since then and the management of Georges fruit and vegetable shop hinted that a major obstacle to working with the farmers is their inaccessibility, especially during rainy periods and this affects their planning and hence impacts negatively on their business. This is because during rainy days, produce from the Foundation Community Project cannot be transported to the market due to poor road conditions. Of course, this leads to disruptions at the farm level when harvesting schedules are affected. On the side of the farm project, this situation has negative consequences on the project's cash flow. Again, from the point of view of the Fruit and Vegetable Shop, although the quality of the produce from the project is usually good, it is not always guaranteed. At times it is difficult to commit a truck to travel to the farm to procure from smallholders because information given about the product quality is often unreliable and this raises transaction costs unnecessarily. Consistent quantity is also often a problem and not always guaranteed. The experience of the Georges shop management was that the first two or so trips might yield acceptable quantities and quality, but thereafter it is difficult to get good quality and quantity from a specific smallholder farmer. The farmers might then go for three months without any marketable produce and hence such an erratic production and supply pattern is not acceptable in the fresh produce business where consumers need the products on a daily basis.

9.6 Chain activities

There are definite activities that constitute governance of supply chains in the generally accepted sense of the term. These are activities that contribute in coordinating the different aspects of the system and direct them to achieve the desired chain goals. But contrary to the implication of the term activity, these are not strictly in the physical dimension, but embrace rules and standards as well as special relational arrangements that help a system to carry out its functions in the way that the stakeholders consider optimal. Within that broad conception, the different types of production and marketing channel governance structures utilised by the smallholders in the municipality have been identified to include:

- Traditional contractual arrangements that oblige households to distribute surpluses in exchange for prior favours – many households report distribution of surpluses outside formal markets.
- A barter system that operates (related to the above) in a situation where access to cash is often limited.
- Formal intermediation through middlemen or bulking agents who visit remote villages in their bakkies to buy surpluses that producers are unable to transport to town centres due to high transport costs, poor rural road links and information asymmetries. It also includes the regulation of activities of chain participants by officials to ensure compliance with standards and rules designed to achieve certain goals for food security, health, and similar aspirations.
- Group formation for production and marketing purposes – in this respect, the associations for collective resource use such as Water User Associations (WUA) or the less sophisticated ones that are limited to single irrigation projects and involves the association of cooperating farmers to rationalise the collective use of resources in order to promote the welfare of members.

9.6.1 Traditional contractual arrangements

Local people in the communal areas have developed mechanisms that allow them to cope with situations of relative scarcity and shortages of basic necessities. In good years, farmers distribute their surpluses to extended family members, friends and neighbours in the hope of the reciprocation of such gestures when the roles are reversed. These transactions were found to constitute important components of the overall transactions that the households engage in and proceed in a very predictable pattern.

9.6.2 The barter system

This is a system of exchanging agricultural produce between consuming units who are not necessarily related as in the traditional contractual arrangements which works as a support structure for the family. The distinctive feature is that this is a cash-free system but is for all practical purposes no different from any other commercial transaction involving agricultural commodities. Instances of this system were observed in the communal areas and seem to be growing in importance as cash shortages become rampant. In Zimbabwe in the period before the unity government was installed, up to the 2009, widespread application of this system was observed as cash virtually ran out and few goods were available to sustain even the most rudimentary transactions.

9.6.3 Formal intermediation

This is usually a specialised intervention for the control and regulation of economic activity to achieve a pre-determined objective. It embodies a range of production and marketing facilitation functions that are needed to support resource-poor chain participants. Production facilitation includes the implementation of a number of support services designed to strengthen the ability of the farmers to produce and market their produce. The government of South Africa is committed to expanding opportunities for small-scale farmers in the country and integrating them more profitably to the nation's agricultural economy. As part of such efforts, several government bodies are providing, through a number of line ministries, assistance in the form of financial support, infrastructure provision, guidance on crop and livestock varietal selection, land management practices, and marketing, and other support services.

A Comprehensive Agricultural Support Programme (CASP) has been implemented for a number of years now and has achieved significant success in addressing the technical constraints farmers face in this era of agricultural restructuring. Another support service specifically dedicated to relieving financial constraints is the MAFISA which is also managed by the Department of Agriculture, Food and Forestry (DAFF). Recently, the Department of Rural Development and Land Reform (DRDLR) has launched a programme to strengthen small-scale and emerging black agricultural entrepreneurs by attaching them to strategic partners who are either established white commercial farmers with proven success in the particular area of agricultural activity, or agencies with expertise in providing technical assistance to the farm sector.

The other type of intermediation is limited to relieving farmers of marketing constraints. This is the function of itinerant middlemen and bulking agents who travel to the remote communal areas and buy farmers' small surpluses for delivery to district or provincial produce markets from where they are subsequently distributed to retail outlets around the province. The value of this intervention is immense because it allows farmers to access markets even when they are severely constrained by high transport costs, lack of transport facilities, and lack of know-how about how to function in modern market environments.

9.6.4 Group formation

Collective action has always been the choice option for relieving resource constraint among resource-poor farmers and other producers and market participants by pooling resources to attain a critical mass of capital base to support profitable economic participation. In addition to being a platform to raise capital, it also affords participants the opportunity to regulate the use of scarce resources and ensuring their more optimal utilisation consistent with the goal of efficiency and profit maximisation. A study was conducted in two communities of the former Transkei homeland area during 2008 and 2009 to assess the impact of collective action on water use patterns, looking specifically at the means for collection of the water, whether or not there was some regulation and control on the use of the water, and what crop enterprises the water was being used for. Important differences were observed between the project and non-project categories (Muchara *et al.*, 2010). The project category represented an expression of collective action while the non-project mode represented individual operations. What emerged was that the non-project water use was less well organised, with community members generally collecting water by means of assorted receptacles such as buckets with which the water is then carried to the garden for watering of the crops. This was clearly a very rigorous and laborious process that could only be undertaken on a very small scale. Thus, despite the absence of regulation and control, the community members were compelled to use very little water that obviously fell below the water requirements of the crops and livestock. It was also observed that under individual operations, the households concentrated exclusively on maize production.

On the other hand, the projects in both communities seemed better organised in respect to water use and systems were in place to promote more optimal water use. For instance, a diesel pump was installed in a regularly maintained irrigation scheme which catered to both crops and livestock. There was also greater diversity in the crop mix, with the projects growing butternuts, cabbages, cauliflower, broccoli, potatoes, green pepper and spinach in addition to maize. Unlike the individual operations, the group projects generated surpluses that were sold in the local market and beyond. There was also the perception among the farmers that the project had made significant improvements in their living conditions and they were earning more income than previously.

9.7 Conclusion

The primary aim of this chapter was to gain some understanding of how the smallholder producers in the Eastern Cape Province are linked to their relevant local and international supply chains and the processes that move the food along the relevant food systems to achieve the goals of the various chain participants. To address these broad question, a set of specific questions were posed to derive a wide range of information on the needs and aspirations of the various participants, sources and destinations of principal foods, nature and patterns of transactions undertaken, coordination and enforcement mechanisms in place, types and levels of municipal regulatory systems implemented, forms of cooperation and group structures in place, and the extent of links with structures and institutions outside the province and South Africa.

The clear indication emerging from this investigation is that substantial variability exists in the product mixes, the participant profiles, existing linkages and mechanisms for coordination and regulation of the food systems and the collaborative mechanisms adopted by participants to cope and adapt with the current and prospective configuration of the system. Without a doubt, supply chain effectiveness remains a central concern for policy makers and producers. The strong evidence that governance modes are important determinants of these provides a solid basis for programming support services to improve smallholder livelihoods. Some conclusions that can be drawn in this regard relate to the role of the human dimension in economic production, the resources and opportunities in the food supply chains, the composition of the food system in terms of the commodity mixes and diversity of participants, and the coordination mechanisms.

In respect to the human dimension, the principal issue is the nature of the needs and aspirations and the fundamental values that drive economic participation in any society. Whether or not a particular governance mode is appropriate depends on the unique production relations in the area and which in turn depends on the unique definition of the goal of production by the participants. People define their own goals depending on their needs and then produce in line with those aspirations. Accordingly, arrangements to manage, coordinate and regulate the production and distribution processes will depend on the purpose of production and distribution consistent with the locally-defined goals and aspirations. Evidently, a subsistence-oriented system will behave differently from a market-oriented system. Structures for coordination and regulation will be selected and applied according to their relevance to the production and distribution goals of the particular environment under consideration. The observation that smallholder production is, for the most part, subsistence-oriented means that the participants and the system are completely different from what obtains in the commercial, market-oriented systems. The governance structures suitable for such a system will no doubt be those that allow for transactions conducted outside the market and depend on rules that are appropriate for non-market situations. Expectedly, more informal arrangements and mechanisms may be prevalent and alignment of such arrangements with formal support structures may be more difficult to achieve. There are possibilities that such a situation may explain the failure of on-going policies to make substantive changes in the pace of reform in the agricultural sector with respect to the smallholder production. While public sector support services are designed to direct production at reasonably medium to large scales, the smallholder realities make it inappropriate to apply such structures.

The other important element of the food system which has important practical implications for the nature of the governance arrangement is the resource profile and opportunities in the food system. The more diverse the resource profile and the more sophisticated it becomes, the more complicated the requirements for governance in a way that meets the goals of production and distribution. What emerged from the investigations is that the local communities continue to suffer serious resource constraints which limit access to the considerable opportunities in the area. For instance, it was observed that transport and access infrastructure are in short supply and the rural areas in the former independent homelands suffer severe infrastructure deficiencies. The demand on systems for coordination are therefore considerable in this respect because special arrangements are often required to deal with even the most simple relationships.

The other important question addressed was the composition of the food system. The study identified a diverse range of crops and livestock in the system. The clear indication is that the farming system is clearly responsive to local dietary needs in terms of the commodities that are produced and distributed. Such diversity imposes some obligation for coordination. The arable crop plots that are located close to the homesteads are particularly vulnerable to damage by itinerant livestock such as sheep and goats but also some cattle. It was found that the traditional leadership plays an important role in this case by establishing guidelines for grazing of communal pastures and at what times of the day or year the domesticated animals are allowed to roam and under what conditions. It was clear that the local arrangement for controlling the use of communal land works very well although occasional conflicts do arise.

The study also evaluated the broader existing coordination and regulatory mechanisms and the sorts of linkages prevailing between the domestic participants and institutions and arrangements beyond provincial and national boundaries. The clear indication is that multiple arrangements are in place. However, the majority of these seem inappropriate to the circumstances of the smallholder producers and small-scale retailers and hawkers operating in the remote rural locations and small-sized urban areas or towns. The local municipalities have units at the administrative headquarters that deal with the regulation of business and other activities within the municipality. These include business registration and licensing, monitoring of business practices in a way that is consistent with the broader provincial and national laws and norms, and administration of incentives of various forms to either encourage or discourage economic participation of one type or another. Some forms of regulation are undertaken by the traditional leadership for activities that are restricted to the communal areas. These are mainly in primary agriculture and in the retailing of basic staples within the communities and around them.

The other form of supply chain governance observed in the district has to do with the quality control in the food system. The Fruit & Veg Hypermarket in King Williams Town revealed the elements of this sort of governance in the relationship it maintains with the small-scale vegetable producers that supply it with produce. The Product Manager indicated that the supermarket chain has a policy to procure produce from local producers up to at least 20% of the shop's turnover. Unfortunately, the local producers targeted under this arrangement hardly meet the requirement due to their low scale of operation.

References

Bembridge, T.J., 1986. Characteristics of progressive small-scale farmers in Transkei. Social Dynamics 12: 77-85.

Chikazunga, D., A. Louw, L. Ndanga and E. Biénab, 2007. Smallholder farmers' participation in restructuring food markets: the tomato subsector in South Africa. Regoverning markets: small-scale producers in modern agrifood markets. University of Pretoria, Pretoria, South Africa. Available at www.regoverningmarkets.org.

Gwynne, R.N., 2006. Governance and the wine commodity chain: upstream and downstream strategies in New Zealand and Chilean wine firms. Asian Pacific View Point 47: 381-395.

Hebinck, P. and P. Lent, 2007. Livelihoods and landscapes: the people of Guquka and Koloni and their resources. Brill Publishers, Leiden, the Netherlands.

Humphrey, J, 2005. Shaping value chains for development: global value chains in agribusiness. Deutsche Gesellschaft fur Technische Zusammernarbeit (GTZ) GmbH, Eschborn, Germany.

Kirsten, J. and W. Moldenhauer, 2006. Measurement and analysis of rural household income in a rural household income in a dualistic economy: the case of South Africa. Agrekon 45: 60-77.

Mather, C., 2004. Regulating South Africa's citrus commodity chain(s) after liberalization. School of Geography, Archaeology and Environmental Studies, University of the Witswatersrand, Johannesburg, South Africa. Available at: http//www.tips.org.za/node/211.

Mdaka, B., 2010. Governance, policies, and programmes promoting public participation at local level and their impact on rural development. Presidential Hotline, Department of Rural Development and Land Reform, Pretoria, South Africa.

Mnduze, B.S. and S.M. Njajula, 2007. Impact of contract farming on small scale farmers: a case study of Zanyokwe irrigation scheme. Mimeo. Department of Agricultural Economics and Extension, University of Fort Hare, Alice, South Africa.

Mokitimi, N., 2007. Institutional constraints to horticulture production and marketing in Lesotho. Paper Presented at the Workshop for Inception of the Fund for Regional Innovative Collaborative Project (FIRCOP. Sept 2007, Johannesburg, South Africa.

Muchara, B., B. Letty, A. Obi and P. Masika, 2010. Deliverable 6: value chain analysis report – identification, mapping and empirical investigation of appropriate food value chains in relation to water as a production input by subsistence and emerging farmers at Willowvale Project Site. Water Research Commission Project K5/1879/4, University of Fort Hare and Institute of Natural Resources, Alice, South Africa.

Muchara, B., 2011. Analysis of food value chains in smallholder crop and livestock enterprises in Eastern Cape province of South Africa. MSc thesis. Department of Agricultural Economics & Extension, University of Fort Hare, Alice, South Africa.

Mzamo, N.C., T.P. Nywebeni and F. Peter, 2007. gender differences in agricultural productivity. Mimeo. Department of Agricultural Economics and Extension, University of Fort Hare, Alice, South Africa.

Ngxetwane, V., 2008. An evaluation of the importance of livestock in the communal areas of Eastern Cape: a case study of Nkonkobe and Amahlathi municipalities. Mimeo. Department of Agricultural Economics and Extension, University of Fort Hare, Alice, South Africa.

Obi, A. (ed.), 2011. Institutional constraints to small farmer development in Southern Africa. Wageningen Academic Publishers, Wageningen, the Netherlands.
Oettle, N., S. Fakir, W. Wentzel, S. Giddings and M. Whiteside, 1998. Encouraging sustainable smallholder agriculture in South Africa. Social Dynamics 12: 77-85.
Umhlaba Consulting Group, 2006. Feasibility study and sustainability plan for the emerging Alice-Kat citrus farmers. Amathole District Municipality, South Africa.
United Nations General Assembly, 2010. Stronger development partnership, better market access, improved governance. Press Briefing, United Nations, New York, NY, USA.
United States Agency for International Development (USAID). 2006. The value chain framework. Briefing Paper, USAID, Washington, DC, USA.
Water Research Commission (WRC), 2005. Sustainable techniques and practices for water harvesting and conservation and their effective application in resource-poor agricultural production – a situational analysis of Guquka and Khayalethu at Nkonkobe Municipality Eastern Cape. Water Research Commission, Pretoria, South Africa.
Williamson, O.E., 2009. Transaction cost economics: the natural progression. 2009 Nobel Prize Lecture, Oslo, Sweden.

10. Smallholder market access and governance in supply chains

Aad van Tilburg, Litha Magingxa, Emma V. Kambewa, Herman D. van Schalkwyk and Alemu Zeruhin Gudeta

10.1 Background

Smallholders in developing countries tend to sell their products at local markets because of their proximity and the fact that they are immediately paid for the produce delivered. Increasingly, they also perceive opportunities in both national or international markets or supply chains to sell their surpluses. They tend, however, to encounter several constraints. A first challenge regards the availability and accessibility of resources and competences which are required to deliver the products that consumers demand (Ingenbleek and Van Tilburg, 2009). Another challenge is the manner in which farmers can be organised to meet the quantities and qualities that their supply chain partners or the consumer market need. A third challenge regards the limited access smallholders have to market information and necessary services such as working capital to manage their operations properly.

The focus of this chapter is the assessment of the extent to which smallholders have access to supply chains in developing countries and Africa in particular, and how this is related to the governance mode of the supply chain. The second section discusses the framework of analysis. In Section 3 the two case studies are introduced and discussed, whereas in Section 4 lessons learnt are discussed by linking the case study material to the theoretical framework presented.

10.2 Framework of analysis

There are structural constraints facing smallholder farmers in many developing countries which are embodied in high transaction costs for information, contract negotiation or contract enforcement resulting in barriers to market access. Delgado (1999) postulates that smallholders require improvements in access to assets, information, services and remunerative markets – implying overcoming high transaction costs – if they are to contribute effectively to economic growth (Table 10.1).

In Table 10.2, we link the factors in Table 10.1 to three governance modes in the supply chain as presented by Delgado (1999). Delgado distinguished three main forms of vertical integration: the independent smallholder operators (IS), small operators linked by contract to processors or marketers (CF) and large commercial operators that tend to be specialised and somewhat vertically integrated (LF). The most predominant form of production in the region is the category of independent operators (IS). They buy their inputs and sell their produce independently through whatever options are available to them. Typical arrangements for the second category, the small operators linked by contract to processors

Table 10.1. Four keys to increase smallholder market participation in Sub-Saharan Africa and four organisational requirements (adapted from Delgado, 1999).

Keys	Need for institutions to
• Access to assets: human, social and economic capital	Implement a net asset transfer to smallholders that provides an incentive for increased productivity
• Access to information to improve both smallholders' production and marketing	Overcome the principal-agent problems in sharing production and marketing information
• Access to services, e.g. farm inputs, contractor services, financial services, transport services	Share the risks of service delivery to smallholders; overcome economies of scale in production
• Access to remunerative markets: how to bridge the gap between production and marketing opportunities?	Overcome economies of scale and quality problems

Table 10.2. Key factors for smallholder market participation by type of governance mode (adapted from Delgado, 1999).

Access to	Independent small operators	Contracts between small operators and processors/marketers	Vertically integrated large farms or plantations
Assets			
Information			
Services			
Remunerative markets			

or marketers (CF) are contract farming or fishing, producer-cooperatives and outgrower schemes with a nucleus estate or large commercial farms. The farmer or fisherman typically gains the benefit of assured supplies of the right inputs at the right time, credit against harvest or catch delivery, and an assured output market. The third form of producer organisation involves large commercially oriented private operators that have some form of vertical link with processors or export marketers (LF).

These types of arrangements are found both in the farming and fisheries sectors. In the latter sector fishers tend to be contracted to large scale fishers. Sometimes these are contracted such that large scale fishers provide fishing gear to fishers who in turn get commission on the basis of quantity of fish caught. In that way the large scale buyers provide linkages with export markets.

Delgado (1999, Table 10.2) reviewed the factors in rural Africa most likely to be associated with transaction costs and how these shape the type of producer organisations most suited in dealing with them. Many of these factors such as economies of scale, returns to extension, required inputs, investment requirements, quality specificity, perishability and distant markets, favour some form of vertical integration such as contract farming or vertically integrated commercial farms or plantations, contrary to the small independent operators.

After defining smallholder market participation or market access, we continue the discussion by presenting factors that influence different types of market access for different governance modes.

10.2.1 Market access

Market access refers to the processes by which people access markets in relation to the nature, efficiency and costs of these processes (Killick *et al.,* 2000). Market access can be related: firstly, to information about product availability, attributes and prices, including the frequency, quality and cost of this information; secondly, to information about counterparties to transactions, as trustworthiness is critical if payment is not instantaneous or checking of quality is costly; thirdly, to the extent that suppliers can have confidence in market conduct; and fourthly, to information on physical costs of accessing a market which is a function of the quality of infrastructure and the transport sector; and finally, to the actual price levels found in the markets in which people transact. The above definition is close to the IFAD (2004) definition, where market access is related to three dimensions *viz.*, physical access to markets (distances, costs, etc.); structure of the markets (asymmetry of power relations between farmers, market intermediaries and consumers); and the level of producers' human capital (e.g. understanding of market forces, prices, bargaining, etc.).

Personal competences

Smallholders need to have sufficient competences in the domain of acquiring and improving their human, social and economic capital base including the traditional tripod land, labour and capital to be able to participate with success in supply chains aiming to produce good quality products for selected markets (see e.g. Nel *et al.*, 2007). Empowerment of smallholders through education of general skills and access to vocational training, extension services and mentorship programs is crucial. Institutions that can be instrumental in this respect can be extension services by public institutions and contract partners, vocational training, farmer study groups, market newsletter services, etc. In addition, special interest groups may influence empowerment of smallholders, e.g. National African Farmers' Union (NAFU), World Wildlife Fund (WWF) and others.

Matching differences in scale and quality

The quantities and qualities produced by individual smallholders seldom match the requirements of subsequent actors in the value chain (Ruben *et al.*, 2007). Activities like grading, sorting and bulking aim to match the requirements of the next stage in the value chain, e.g. the processing industry or the wholesale sector in case processing is not required. The question is which party or parties are taking the coordination role. Are smallholders willing and able to perform a part of the roles of coordinating and value adding through group action by, for example, a cooperative packing station or auction? Alternatively, is it the demand side of the market that is coordinating supply, for example, through assembly in rural markets, contract farming, or outgrower schemes connected to plantation agriculture?

Access to information

Smallholder farmers in sub-Saharan Africa face a range of marketing and exchange problems, among which informational constraints are much cited. In their dealings with the market, smallholder farmers find themselves at a major disadvantage. Many do not understand how markets operate and why prices fluctuate; they have little or no information on market conditions and prices; they are not organised collectively; and they have no or limited experience of market negotiation (Freeman and Silim, 2001; Heinemann, 2002; IITA, 2001;). Producers experience a weak bargaining position *vis-à-vis* traders because often they do not have timely access to salient and accurate information on prices, locations of effective demand, preferred quality characteristics of horticultural produce, nor on alternative marketing channels. Increasingly, mobile phone and internet technology tend to improve smallholders' access to information.

Access to services

Liberalised markets do not necessarily attract small producers to a significant extent unless specific support measures are taken. There are a number of 'invisible', yet critical, barriers to trade that must be overcome, such as the lack of awareness of market opportunities and of familiarity with standards, limited scale of operations and specific skills, among others (IFAD, 2004).

10.2.2 Governance mode of a supply chain

This section discusses modes of chain governance and factors or institutions that may affect the linkage of smallholders to supply chains.

Chain coordination

In lengthy supply chains, some form of coordination is required to match demand of final customers with supply. In principle, coordination in the value chain can be spot market-dominated where market information and market prices are the coordinating devices; it can be hierarchy-dominated through ownership or with contracts as coordinating devices, or it can be network-dominated where personal, trust-based relationships are the coordinating device (Fafchamps, 2004; Powell, 1991; Van Tilburg *et al.*, 2007). Individual smallholders who are not involved in value chains tend to sell in oligopsonic markets characterised by many suppliers and a few buyers, which generally results in a weak negotiation position for these smallholders (Ruben *et al.*, 2007). Their countervailing power can be increased considerably through group action in both the procurement of inputs and the market supply of products. In the end, smallholders may be able to change the market structure to their own advantage when subsequent value adding activities are integrated in one and the same value chain.

Channel decisions in hierarchy-dominated supply chains tend to be taken by a channel leader with the aim of serving the customers better than would probably be the case if the products were sold at anonymous (spot) markets. Several 'prototypes' of supply chains have been distinguished in the literature (e.g. Gereffi, 2005). Stern *et al.* (1996) considered (1) Conventional marketing channels 'consisting of isolated and autonomous units or stages, each of which performs a traditionally defined set of marketing functions. Coordination among channel members is primarily achieved through bargaining and negotiation at spot markets'; (2) Vertical marketing channels 'consisting of networks designed to achieve technological, managerial and promotional economies through the integration, coordination, and synchronisation of marketing flows from points of production to points of ultimate use'. Vertical marketing systems are coordinated by joint planning, contracts or corporate ownership; (3) Networks of agents based on trust relationships, e.g. among relatives or people belonging to the same group or clan that shares particular values; and (4) Hybrid forms of governance being a combination of at least two of the previously mentioned types. Usually, there is no 'one form fits all' solution for coordination of a supply chain and, consequently, different parts of a supply chain may have different governance modes. Supply chain leaders may improve the value added in the supply chain by focusing activities on a few particular market segments to be served or the needs of a narrowly defined market niche. The supply chain may be instrumental in escaping the commodity trap by transforming (homogenous) products into unique or specialty products which may be further articulated by labelling or branding, e.g. the case of Rooibos tea.

Choice of governance mode

Modern market-oriented chains tend to become shorter as intermediaries between producers and parties downstream in the chain become superfluous because of better information and vertical coordination. Inter-company relationships in these chains tend to be influenced by (transaction-specific) investments of processors or exporters (e.g. cold stores and the provision of inputs: seeds, pesticides, credit) to minimise delivery uncertainty and increase the consistency in quantity and quality of deliveries (Ruben *et al.*, 2007). Transaction-specific investments require more certainty about the long-term relationship between trade partners. The physical distance between producers and consumers of food products ('food miles') is generally large in the case of global supply chains implying that actors at both ends of the chain tend are not readily aware of each other's needs, opportunities and constraints. Consumers might not be aware of the conditions under which producers have to work, and producers may not understand legitimate concerns of (segments of) consumers. For example, with respect to sustainable resource use (land degradation, pollution of the environment and reduction of biodiversity); social responsibility issues (e.g. abuse of the labour force or child labour); or food quality and safety (which may be important to various groups of stakeholders including buyers, suppliers and both internal and external institutions) (Ruben *et al.*, 2007). Generally, a hierarchical governance mode tends to be more suitable for perishable products with variation in market demand than for non-perishable products offered in stable markets.

Power distribution in the supply chain

To illustrate this discussion we will compare two governance modes: (a) an individual smallholder concluding a contract with a buyer in the supply chain, and (b) a joint action of smallholders by concluding one contract as a group with a supply chain buyer. In the first case, the required coordination in the supply chain is carried out by the buyer. However, in the second case, the coordination task can be shared between the group of smallholders and the buyer. Examples of the first case regard individual barley growers who conclude contracts with a large malt manufacturer, or dairy farmers concluding a contract with a dairy firm or group. Examples of the second case are cooperatives of smallholders concluding a contract with a processing industry for the delivery of an agricultural commodity or a cooperative of herbal tea growers concluding a contract with a processing company (Nel *et al.*, 2007).

Response to supply chain dynamics

It is essential for (groups of) smallholders participating in a supply chain to be able to respond to chain dynamics with respect to changes in quantities or qualities demanded, e.g. demand for product upgrading, changes in bargaining conditions due to changes in the market structure, or proposed changes in the governance mode of a value chain. Examples regard variation in required consumer products, changes in type and quality of market information and changes in national or international regulations on product safety or trade. These changes may affect both the primary supplier's position in the value chain and the distribution of value added among supply chain partners (Gibbon, 2003; Nel *et al.*,

2007; Ruben *et al.*, 2007). Relevant institutions in this respect are, for example, market information services such as press releases, trade journals and specialist workshops.

In summary, smallholders need both their skills and support from chain actors and the institutional environment in order to be successful as partners in the value chain. The foregoing considerations are summarised in Table 10.3.

10.3 Case studies on domestic and international distribution

10.3.1 Domestic distribution

This case study regards institutional development for improved smallholder market access in South Africa. The case examines the critical elements of the institutional framework necessary for improved smallholder market access (Magingxa 2006).

Introduction

There is growing evidence that many smallholder farmers can benefit from market-oriented agriculture. However, high transaction costs are a barrier to accessing the markets. Poor households are often ill-equipped to respond to rapidly changing market conditions and, in some cases, have seen old production strategies undermined by new competition without being able to take advantage of the new opportunities provided by liberalisation policies, e.g. the deregulation of agricultural markets in South Africa. The response of smallholder agricultural production to marketing liberalisation has been mixed.

Small-scale agriculture is extremely important in achieving the government's development objectives because it directly contributes to household food security through meeting subsistence requirements. Smallholder agriculture also has the potential to drive economic growth because medium sized farms are typically more efficient producers than large farms in low income countries (Heltberg, 1998), and have better consumption and investment patterns for stimulating growth in the non-farm economy (Hazell and Roell, 1983).

In spite of the potential that small-scale agriculture possesses, there has been a reverse process in relation to smallholder farming in South Africa. Compared to other countries, smallholders in South Africa are faced with a very large set of constraints. Productivity levels have dropped drastically, especially in the formerly state-managed smallholder irrigation schemes. This is related to declining government support to smallholders in the so-called Irrigation Management Transfer (IMT) process. In Limpopo Province, Kamara *et al.* (2002) noted that the level of production had dropped to about 20% in schemes previously managed by the Agricultural Rural and Development Corporation (ARDC). This contrasted against a general growth in the country's agriculture. They also found that some farmers were producing at a loss. Making particular reference to the Eastern Cape, Bembridge (2000)

Table 10.3. Factors influencing market access by type of governance mode in the value chain.

Access to	Independent small operators	Vertical coordination by contracts between small operators and processors/marketers	Vertically integrated large farms or plantations
Assets			
• Improving personal competences	Inadequate	Included in the terms of the contract	Dependent on the rules and regulations in the value chain
• Matching differences in scale and quality	By the assembly trade	Included in the terms of the contract	Same
• Physical costs of access	Dependent on distance and transport availability	Part of the contract	Same
Information			
• Sharing production and marketing information	Usually not	Included in the terms of contract	Dependent on the rules and regulations in the value chain
• Sharing information about the reliability of contract partners.	Depends on social capital and mutual trust	Concluding a contract implies a certain level of mutual trust	Same
Services			
• Sharing the risks of service delivery	No	Part of the contract	Dependent on the rules and regulations in the value chain
• Response to supply chain dynamics in demand	Dependent on suppliers' opportunities and constraints	Can be part of the contract	Same
Remunerative markets			
• How well are markets regulated?	Depends	By contract	Dependent on the rules and regulations in the value chain
• Overcome economies of scale and quality problems	By the assembly trade	Through proper planning of the contract partners	Same

argues that in spite of huge investments, the performance of most small-scale irrigation schemes in the Eastern Cape has been poor and falls short of the expectations of engineers, politicians, development agencies and the participants themselves.

The large extent of historical dependency in South African small-scale agriculture presents a situation where farmers – especially those based in the large former homeland irrigation schemes – had become accustomed to the profound support provided by parastatal organisations, which managed most of the smallholder irrigation schemes in the country.

Several recent studies confirmed that market access is one of the most critical factors that determine the success of smallholder farming (e.g. Gabre-Madhin and Haggblade, 2001 and Hau and Von Oppen, 2002). It is also generally agreed that access to markets is the critical link to profitability of smallholder farming. In the envisaged role of smallholder farming as one of the vehicles for economic growth, the role of market access is implicit in the message. Magingxa (2006) also concluded that market access is a prerequisite for agricultural commercialisation. The same study concluded that access to markets for smallholder irrigators is largely determined by institutional conditions.

If agricultural reforms in Africa are to fulfill the high expectations of their proponents, improvement will have to be made in four areas (Kherallah *et al.* 2002). First, the task of liberalising agricultural markets must be completed. This task may imply the withdrawal of state enterprises from direct agricultural production, marketing, and processing, as well as convincing signals from political authorities that the reforms will not be reversed or undermined. Second, complementary policies in other sectors are needed to enhance the benefits of the reforms and alleviate the negative effects. A stable macro-economic environment, progress in taming corruption, and stronger legal infrastructure are prerequisites for stimulating domestic and international investment, including that in the agricultural sector. Similarly, programmes to provide a credible safety net for households adversely affected by the reforms are justifiable on their own terms as well as for the political sustainability of the reforms. Third, the withdrawal of the state from commercial activities should not be interpreted as withdrawal from its essential role in providing public goods. Governments and international organisations need to reverse declining investments in agricultural research and extension, improve transport infrastructure, promote the sustainable use of natural resources, and develop public services such as market information, plant protection, and disease control. Fourth, the government can play a role in promoting non-governmental institutions in the agricultural sector. Farmer associations facilitate dialogue between the government, on the one hand, and farmers and traders on the other. This dialogue should guide the design of public institutions such as grades and standards, plant protection regulations, and market information services.

The dual nature of the agricultural sector in South Africa, consisting of a well-established commercial sub-sector and a small-scale sub-sector, provides a contrasting picture and requires specific consideration. The established commercial sector and the areas in which commercial agriculture preponderates are served by a sophisticated agricultural marketing system with infrastructure supporting agricultural production and marketing (Van Schalkwyk *et al.*, 2003). Development can hardly be expected and no uplifting of the rural poor can occur in the absence of significant improvements in the provision of marketing services serving particularly the small-scale producers in the rural areas. Generally, to overcome these problems, farming communities have formed cooperatives, collective marketing associations, and other mutual alliances to increase their buying and selling power in the market place. Larger commercial players have also been active, forming mutually beneficial alliances with farmers supplying marketable products at agreed prices. Clearly, it is only by such means that most developing country farmers can move from a poverty cycle to an income cycle, and begin to make a real contribution to overall economic development (IITA, 2001). Other options explored in literature include warehouse receipt systems (Coulter and Onumah, 2002), contract farming (Kirsten and Sartorius, 2002), and a rural assembling point system (Freeman and Silim, 2001).

The information provided is largely based on a case study, which forms part of a larger study on smallholder market access and supporting institutions in South Africa (Magingxa, 2006). A brief account of the methodology and results from that study is provided. Components of a response framework to the problem of smallholder market access are presented, based both on relevant literature and results from this South African case study.

Summary of the case study

Data and methods

The study was conducted in six smallholder irrigation schemes in three provinces of South Africa: Eastern Cape (Melani, Qamdobowa, Roxeni and Somgxada), Limpopo (Sepitsi) and Mpumalanga (Hereford). The Eastern Cape and Limpopo provinces contain most of the smallholder irrigation schemes in South Africa. These are also classified as the poorest provinces in the country (SSA, 2001). The selection of schemes was conducted taking the requirements of the funding institution into consideration and ensuring representativity of the sample. Market access problems are also very evident within the smallholder farming areas selected. In all the selected study areas, market access was cited as a key impediment to the farmers achieving their income goals.

Based on a review of the literature, 14 explanatory variables were selected addressing critical aspects of market access – the dependent variable – in smallholder farming. The selected variables involve: infrastructure, information availability, credit availability, extension availability, market association, transport, transport costs, training, creativity, entrepreneurship, distances, farmer age and off-farm income.

Principal component analysis was performed on the explanatory variables to address multicollinearity. The analysis yielded six components interpreted as physical access to market (PC_1), farmer skills (PC_2), nature of access to market (PC_3), inventory of support services (PC_4), ability to respond to opportunities (PC_5) as well as off-farm income (PC_6). A logistic regression analysis was conducted to determine the influence of these principal components on the variable to be explained,. Simulations were carried out to determine the effect of variation in each of the significant principal components on market access.

Results and discussion

Results of the regression analysis, as shown in Table 10.4, revealed the most significant components as physical access to market, farmer skills and nature of access to market.

The model has a 82% goodness of fit. PC_1 and PC_3 are significant at the 1% level while PC_2 is significant at the 5% level. PC_1 represents physical access to the market and is highly significant. Physical access to markets is a function of the quality of infrastructure and the organisation of the transport sector. Transport cost has always been shown as the single biggest source for the cost of accessing the market for smallholders. Omamo (1998) argues that the cost of transport affects crop choices of smallholder farmers and this leads to cash crops being forgone for less profitable crops. The outcome of this research therefore confirms that.

PC_2 represents farmer skills. It is a result of positive interactions between creativity and entrepreneurship. The fact that it has come out as significant could be tied to the prevailing agricultural marketing situation in South Africa. Until recently, up to 1996, marketing of most smallholder produce was done mainly by parastatal organisations through agricultural

Table 10.4. Logistic regression of six principal components on market access.

Variables	Coefficient	Standard error	T-ratio	Probabilities
Constant	0.528	0.34889	1.5132	0.133
Physical access to market (PC_1)	-2.478	0.71307	-3.4749	0.001***
Farmer skills (PC_2)	0.695	0.29674	2.3417	0.021**
Nature of access to market (PC_3): individually or group action	-1.45	0.34556	-4.2031	0.000***
Inventory of available support services (PC_4)	-0.272	0.21050	-0.12902	0.898
Ability to respond to opportunities (PC_5)	-0.258	0.27400	-0.094009	0.925
Off-farm income (PC_6)	-0.384	0.27702	-1.3852	0.169

*** Significant at $P \leq 0.01$ and ** $P \leq 0.05$.

product control boards. The dissolution of these boards in the wake of liberalisation of agricultural markets calls for creativity and an entrepreneurial spirit on the part of smallholders to continue the marketing of their produce.

PC_3 involves whether markets are accessed individually or as part of an organised grouping. The nature of organisation is becoming an increasingly important aspect for smallholder farmers in developing countries in general with respect to accessing credit and improving the bargaining power at the market.

Subsequent simulations displayed the critical components in smallholder market access in South Africa and demonstrated that a reduction in the physical cost of accessing the market and a reduction in transaction costs associated with the nature of access to the market would lead to a significant increase in the probability of accessing the market. The simulation also showed that an improvement in the farmer's skills and knowledge would lead to an increase in the probability of accessing the market.

Conclusion of the case study on domestic distribution

Clearly, the problem of physical access to market cannot be viewed as divorced from embedded institutional problems in smallholder irrigation management. The lack of proper local markets that are linked to outside buyers where smallholder irrigators can sell their produce is clearly an institutional issue. The market is an institution and therefore needs to be viewed as such. It is where exchange takes place between buyer and seller and in order to function well, it needs rules and regulations that have to be followed by parties involved in the transaction. A lack of profitable local markets is a source of transaction costs for most smallholders.

The argument advanced above can easily be extended to farmer skills. This issue can also be understood in relation to institutions that facilitate farmer creativity and entrepreneurship. Absence of proper institutions to address these, led to the *status quo*. Farmer organisations are well within the amplitude of institutions and therefore warrant an institutional treatment.

In addition to the two aspects addressed above, the nature of access to the market has shown to be critical. Farmers can access markets either as individuals or as part of an organised grouping, e.g. a market association. Organised farmers stand a better chance of accessing the market than individuals (e.g. Heinemann, 2002). This is partly because of the small volumes produced in smallholder operations and partly because when organised members as a grouping, farmers are in a better position to bargain with formal business who rely on assured quantities at agreed times for retailing.

It can be concluded from these observations that there is a need for a new approach in addressing the problem of market access. The conclusions made above have crucial implications for possible steps that can be taken to deal with the problem of smallholder irrigation management.

10.3.2 International distribution

This case study deals with conditions that shape smallholders' participation in international supply chains (Kambewa, 2007).

The relationship between smallholders and international agricultural markets

Development literature continues to argue that globalisation of markets offer opportunities for economic growth especially for developing economies (Bardhan, 2006; Thorbecke and Nissanke, 2006). However, such opportunities and corresponding benefits do not always reach smallholders in developing economies who fail to participate in international supply chains. Participation in international food chains is increasingly dependent on guarantees for *food safety* as a bottom line and *food quality* as a competitive edge (Ruben *et al.*, 2007). Implementing standards and institutions for ensuring food safety and quality require relatively large financial and organisational resources, which are not easily met by emerging farmers (Henson and Loader, 2001; Panisello and Quantick, 2001). Consequently, smallholders are often marginalised from global networks (Van der Meer, 2006). This section synthesises some of the institutional and technical opportunities and limitations for smallholders participating in international supply chains by briefly reviewing: the institutional environment for international trade; food quality and safety regulations from the European Union (EU) perspective; and consumer demands in Europe. Lastly, we present a case study on how in the 1990s the EU food safety regulations affected the Nile perch supply chain from Lake Victoria.

The international institutional environment

The highest level at which global trade regimes are formulated and negotiated is the World Trade Organization (WTO) forum. As Ponte, Raakjaer and Campling (2007) observe, WTO negotiations often present hurdles to developing economies, especially those that tend to have bilateral relationships with developed economies. For example, some African countries enjoy preferential access to the EU as well as Unites States of America (USA) markets through bilateral trade agreements such as African Growth Opportunity Act. Such agreements sometimes bind developing countries when they need to take trade decisions of regional importance. For instance, some African countries allow foreign (e.g. EU) fishing vessels access to their waters and such governments thereby earn significant foreign exchange from the fees paid by the fishing vessels. Such countries may find it difficult to negotiate on terms that may reduce their earnings from foreign fishing vessels. Countries without such bilateral agreements may not always benefit from the WTO negotiations because their demands may be obscured by self-interests of the bilateral agreements. At the same time, countries with bilateral agreements with the West may not always appreciate, for example, that the presence of a foreign fleet ultimately precludes the development of their domestic fishery fleet, which cannot compete with Western fleet in both the waters and markets, leaving domestic fisher folk out of production and markets.

The effect of food safety standards and regulations

Although traditional trade barriers (e.g. tariffs) are steadily being reduced, food safety standards requiring issues such as traceability, product certification and environmental standards are increasing in scope and significance as international food trade opens up (Frohberg *et al.*, 2006). Developing economies and smallholder producers in particular, must comply with increasingly stringent food safety standards to access western markets. The EU, for example, consistently and rapidly changes food safety standards and requirements (see Arvanitoyannis *et al.*, 2005 for a review). The sheer number of regulations and the frequency of amendment mean that smallholders must constantly change their production systems to comply with rapidly changing food safety requirements. Without adequate quality management facilities, compliance is nearly impossible.

The eligibility of developing economies is also subject to additional preliminary conditions. In the meat sector (including fishery products), for example, factors such as the country's exotic diseases and the environmental health situation; the ability to regularly monitor and timely supply information on infectious or contagious diseases; the manner in which the country organises and implements disease preventive and control measures affect the countries eligibility to access to EU markets.

Whereas food safety measures are often enforced through regulatory mechanisms, quality standards take a different form. Global food markets are increasingly being saturated with quality differentiated food products that try to address consumers demand for advanced quality attributes (Ponte and Gibbon, 2005). These evolving quality demands have, over the past decade, heightened the importance of quality attributes that include physical product characteristics and also the emotional, situational, environmental and traceability attributes (Mansfield, 2003). Supply chains develop codes of conduct to communicate and assure consumers of the safety and quality of the products they purchase. Smallholders are affected by these issues but they are in no position to influence them due to the fact that they cannot constitute an effective lobby because of their small scale and limited resources, lack of sufficient access to information, etc.

Food quality standards offer opportunities for the inclusion of smallholders in international markets through provision of information on production standards (e.g. Fulponi, 2006). However, costs associated with compliance lead to exclusion of many smallholders from these markets where standards are important, often require major structural changes in production practices and processes and also require access to information (Frohberg *et al.*, 2006; Humphrey *et al.*, 2004). Smallholders in developing countries cannot cope with the increasing cost of investments, communication and monitoring (World Bank, 2005). Small and medium producers from developing countries rarely comply with food safety and quality standards without support from downstream actors (Chemnitz *et al.*, 2007). Downstream support, however, may not always be beneficial. Smallholder producers may lose control over production and management decisions and sovereignty (Chemnitz *et al.*, 2007). Smallholders compromise their bargaining power (Bardhan, 1980) and in the long term, lose their autonomy. Eventually, the cycle of poverty may never be broken for smallholders.

Changing consumer preferences and demands in Europe

Rapid changes in consumer preferences and demand give another perspective of the external factors affecting smallholder participation in international supply chains. Increasing concerns over food safety, the changing life styles and increasing incomes (Leeflang and Van Raaij, 1995) are influencing the way food is being produced and managed. Over the past decade, the world has witnessed consumers' endless pursuit of variety, year-round supply of fresh produce, 'healthy' food, convenience food, timeliness of supply and value added which has created a market climate where quality attributes serve as the basis for differentiation and market segmentation (Verhallen *et al.*, 2004). The supply chain's competitive advantage depends on the ability of suppliers to deliver the right product mix, when, where and in the form demanded. Quality standards are also governed by rules of collective action by a number of supply chains (e.g. EurepGAP, British Retail Consortium). Smallholders seeking to supply to such retail chain must meet such requirements.

What does compliance with food safety standards mean for smallholders? The case of Nile perch exported from Lake Victoria to Europe

The Nile perch chain from Lake Victoria gives an explicit example of the effect of food safety laws imposed by the EU affect smallholders' entry into international supply chains. Since the mid 1980s, the chain experienced a rapid expansion of export demand especially in the EU which takes up to 80% of the export volume (see www.globefish.org). From mid to late 1990s, the EU imposed three consecutive restrictions on import of Nile perch products on suspicion of food safety threats from *Salmonella*, cholera and fish poisoning that were reported in the chain (Table 10.5). The threats led to import bans for fish and fish products from the region (Henson and Mitullah, 2003). The restrictions were enforced through European Commission Decisions 97/272/EC; 97/458/EC; and 97/878/EC which outlined the conditions and duration of the restriction.

On a positive note, the restrictions triggered investment in the processing stages which upgraded and improved the facilities and started to implement the principles of the Hazard Analysis of the Critical Control Point. The hygiene conditions in the chain improved as

Table 10.5. Food safety restrictions and precautionary measures on Nile perch exports to the EU (Henson and Mitullah, 2003).

Dates	Restrictions/precautionary measure	Cause	Products/regions
27 Nov. 1996 - 3 April 1997	Exports prohibited to Spain and Italy	Spain detected *Salmonella* in a number of Nile perch consignments from Kenya, Tanzania and Uganda	Nile perch
4 April 1997 - 30 June 1998	Border testing of all consignments for *Salmonella*		Nile Perch
23 Dec. 1997 - 30 June 1998	Exports prohibited to EU	Cholera across East Africa largely associated with the heavy rains brought by *El Nino*	Fresh fish
	Border testing of all consignments for Vibrio *cholerae* and Vibrio *parahaemoliticus*		Frozen/processed fish not caught at sea and directly landed to EU
12 April 1999 - 1 Dec. 2000	Exports prohibited to EU	Suspected case of fish poisoning with pesticide in Uganda	Fish from Lake Victoria

processors invested in cold storage and refrigerated transport facilities to transport fish from the landing beaches. On a negative note, some fish processing companies declined as a direct result of the loss of export revenue. The returns from Nile Perch exports during the restriction period were barely sufficient to cover costs. Some processors shed labour or completely suspended operations. The reduced export demand resulted in a decline in the landed price of Nile Perch, reducing the income of fisher folk who were hardest hit (Geheb and Binns, 1997). Whereas the restrictions were justified on the poor performance of the regulatory and monitoring systems, some analysts (e.g. Ponte, 2005) argue that there was no scientific proof that the fish was really 'unsafe' for human consumption. This is in line with those (e.g. Otsuki *et al.*, 2001) that have blamed the food safety standards and regulations as being more trade barriers than measures for protecting EU consumers. The poor fisher folk paid the hardest price: losing their major livelihood sources as a result of the restrictions imposed on the Nile perch chain. Furthermore, the investments (in refrigerated trucks) by processors increased their market power to dictate supply terms, limiting the benefits accruing to the fisherfolk.

10.4 Reflection on the questions posed

The chapter started with the question how smallholder access to markets is related to the governance mode in supply chains. Or, more specific: how are factors of market access such as access to assets, access to market information, access to services and access to remunerative markets related to different types of chain governance? A framework was presented and two case studies have been discussed to illustrate the relevance of the subject. In Table 10.6, we summarise how and to what extent the two presented case studies fit into the framework.

Looking at the cases where the actors were notably independent operators in a static sense, we can conclude that the four factors influencing access to markets appeared to be relevant. But looking at this relationship in a dynamic sense, we expect substantial more vertical integration for at least a part of the smallholder population. The large-scale commercial farmers in South Africa already have a long-standing experience on how to access profitable markets and the rules and regulations in their value chains support them in continuing their profitable business.

An interesting issue in South Africa is how smallholders can benefit from the knowledge and experience of large-scale commercial farms (see Chapter 2). Similarly, access to assets and information in the international distribution example made a difference between success and failure.

Table 10.6. The case studies and factors influencing market access by type of governance mode.

Access to	Independent small operators	Contracts between small operators and processors/marketers	Vertically integrated large farms or plantations
Assets	DD, ID	*DD, ID*	FC
Information	DD, ID	*DD, ID*	FC
Services	DD, ID	*DD, ID*	FC
Remunerative markets	DD, ID	*DD, ID*	FC

DD = domestic distribution and ID = international distribution during the time of the case study; *DD* = domestic distribution and *ID* = international distribution expected for a part of the smallholders in the coming decade; FC = large-scale farmers working mainly for international markets.

10.5 Recommendations for institutional changes

This final section outlines a set of recommendations based on the surveyed literature and results from the study.

Usually, research on smallholder irrigation management is done by research professionals. However, the link between the results of this research and the policy making process is unclear. What is needed here is interaction between policy-makers, scientists and farmers to facilitate a two-way movement of information. To this end, a multi-stakeholder platform where government, private sector, civil society and farmers are partners is recommended. The goal should be to overcome exclusion from policymakers and programme designers to facilitate smallholder market access and private sector service provision. In this regard, the government can play a facilitative role by designing – together with other partners – a framework that will facilitate continued and protected investment by the private sector in rural farming enterprises.

One way of dealing with the problem of high transportation cost may be to introduce incentives for transport contractors involved in smallholder agriculture. Incentives will attract service providers and encourage small farmers to use common transport rather than the more expensive alternative of small pick-up vans. However, that implies that attention has to be paid to the road infrastructure, important for rural farmers to trade with outside buyers because the local market is often not big enough and widely dispersed. This renders it necessary for local, provincial and national government bodies to be aware of the dependence of local rural welfare on High quality transport infrastructure, particularly roads.

There is a need to pay more attention to the issue of capacity building in the small-scale smallholder sector in general. Avenues that can be explored include continued specialised training and encouragement of constant inter-action with more experienced market participants. In this case, government can collaborate with civil society and private sector to facilitate such interaction.

Farmer organisations and associations should be encouraged. A reward system may encourage them. These can serve as security where credit is needed to improve the small-scale irrigation sub-sector. Collective action for smallholders can also improve their bargaining power and move them from the level of price takers to empowered negotiators with marketing agents.

The issue of farmers holding on to land for non-productive purposes can be dealt with in two ways. Firstly, by giving more power to community organisations that should be in a position to distribute productive land to those interested in farming. Secondly, a land market system can be created that will improve the economic value of the land holdings and facilitate exchange. In addition, rural non-farm enterprises may have to be developed that will provide more incentives for committed farmers to remain in the irrigation schemes.

Rebuilding of important smallholder support services and institutions is crucial. Within these, the extension service shows signs of being less efficient in terms of the quality and quantity of interactions. It may be necessary to strengthen this service and encourage specialisation and continued training of extension officers.

Information is cited as one of the major sources of transaction costs. Facilitation of information regarding quality requirements (grades and standards) and prevailing prices in the formal and more lucrative markets is critical. Access to this type of information will assist smallholders to gear up for the opportunities provided. Examples tried elsewhere include riding on the current wave of less costly information technology like use of cell phones, in addition to traditional media.

Developing mechanisms that ensure smallholder access to market information, assets and services cannot be overemphasised. What is crucial, however, is that these should be provided as a package because individually they cannot adequately empower smallholders. In view of the dynamism of the market environment, it is crucial that the mechanisms to improve smallholder access to market and the governance mode they adopt enable them to quickly adapt to changing conditions.

In addition, information exchange within the channel should always take into consideration the competences and abilities of smallholders to comprehend and understand their role and benefits arising from their presence in the value chains.

Whereas the role of farmers organisations is critical to realise economies of scale, the way that these organisations are formed is critical. There are often already informal groups that are formed on a common interest and mechanism to support them need to recognise such groups that have already identified common interest and have structures already in place. Building on existing groups may be more effective than forming new ones that in the end may be seen as external impositions.

References

Arvanitoyannis, I.S., S. Choreftaki and P. PersefoniTserkezou, 2005. An update of EU legislation (Directives and Regulations) on food-related issues (safety, hygiene, packaging, technology, gmos, additives, radiation, labelling): presentation and comments. International Journal of Food Science and Technology 40: 1021-1112.

Bardhan, P.K., 1980. Interlocking factor markets and agrarian development: a review of issues. Oxford Economics Papers, New Series 32 1: 82-98.

Bardhan, P.K., 2006. Globalization and rural poverty. World Development 34: 1393-1404.

Bembridge, T.J., 2000. Guidelines for rehabilitation of small-scale farmer irrigation schemes in South Africa. WRC Report No. 891/1/00. Water Research Commission, Pretoria, South Africa.

Centre for Agricultural and Rural Cooperation ACP-EU (CTA), 2006. Expert consultation on market information systems and agricultural commodities exchanges: strengthening market signals and institutions. Proceedings of an expert meeting, 28-30 November 2005, Amsterdam, the Netherlands.

Chemnitz, C., H. Grethe and U. Kleinwechter, 2007. Quality standards for food products – a particular burden for small producers in developing countries? Paper at the EAAE Seminar Pro-poor Development in Low Income Countries: Food, Agriculture, Trade, and Environment, 25-27 October 2007, Montpellier, France.

Chikazunga, D., A. Louw, L. Ndanga and E. Biénabe, 2007. Smallholder farmers' participation in restructuring food markets: the tomato subsector in South Africa. Regoverning markets: small-scale producers in modern agrifood markets. University of Pretoria, Pretoria, South Africa. Available at: www.regoverningmarkets.org

Coulter, J. and G. Onumah, 2002. The role of warehouse receipt systems in enhanced commodity marketing and rural livelihoods in Africa. Food Policy 27: 319-337.

Delgado, C.L., 1999. Sources of growth in smallholder agriculture in Sub-Saharan Africa: the role of vertical integration of smallholders with processors and marketers of high-value items. Agrekon 38: 165-189.

Dorward, A., J. Kydd and C. Poulton (eds.), 1998. Smallholder cash crop production under market liberalisation: a new institutional economics perspective. CAB International, Wallingford, UK.

Fafchamps, M., 2004. Market institutions in Sub-Saharan Africa: theory and evidence. The MIT Press, Cambridge, MA, USA.

Freeman, H.A. and S.S. Silim, 2001. Commercialisation of smallholder irrigation: the case of horticultural crops in semi-arid areas of Eastern Kenya. In: H. Sally and C.L. Abernethy (eds.) Private irrigation in Sub-Saharan Africa: Proceedings of regional seminar on private sector participation an irrigation expansion in sub-Saharan Africa, 22-26 October, IWMI, FAO and CTA, Accra, Ghana.

Frohberg, K., U. Grote and E. Winter, 2006. EU food safety standards, traceability and other regulations: a growing trade barrier to developing countries' exports? Paper prepared for presentation at the International Association of Agricultural Economists Conference, August 12-18, 2006, Gold Coast, Australia.

Fulponi, L., 2006. Private voluntary standards in the food system: the perspective of major food retailers in OECD countries. Food Policy 31: 1-13.

Gabre-Madhin, E.Z. and S. Haggblade, 2001. Successes in African Agriculture: results of an expert survey. MSSD Discussion Paper 53. International Food Policy Research Institute, Washington, DC, USA.

Geheb, K. and T. Binns, 1997. Fishing farmers or farming fishermen? The quest for household income and nutritional security on the Kenyan shores of Lake Victoria. African Affairs 96: 73-93.

Gereffi, G., J. Humphrey and T. Sturgeon, 2005. The governance of global value chains. Review of International Political Economy 12: 78-104.

Gibbon, P., 2003. Commodities, donors, value-chain analysis and upgrading. Danish Institute for International Studies, Copenhagen, Denmark.

Hau, A.M. and M. Von Oppen, 2002. Market access and agricultural productivity in DoiInthanon villages. International Symposium: Sustaining Food Security and Managing Natural Resources in Southeast Asia – Challenges for the 21st Century – 8-11 January 2002, Chiang Mai, Thailand.

Hazell, P.B.R. and A. Roell, 1983. Rural growth linkages: household expenditure patterns in Malaysia and Nigeria. IFPRI Research Report 41. International Food Policy Research Institute, Washington, DC, USA.

Hebinck, P. and P. Lent, 2007. Livelihoods and landscapes: the people of Guquka and Koloni and their resources. Brill Publishers, Leiden, the Netherlands.

Heinemann, E., 2002. The role and limitations of producer associations. European Forum for Rural Development Cooperation. 4 September, Montpellier, France.

Heltberg, R., 1998. Rural market imperfections and the farm size-productivity relationship: evidence from Pakistan. World Development 26: 1807-1826.

Henson, S. and R. Loader, 2001. Barriers to agricultural exports from developing countries: the role of sanitary and phytosanitary requirements. World Development 29: 85-102.

Henson, S. and W. Mitullah, 2003. Kenyan exports of Nile perch: impact of food safety standards on an export-oriented supply chain. University of Guelph, Guelph, Canada.

Humphrey J., N. McCulloch and M. Ota, 2004. The impact of European market changes on employment in the Kenyan horticulture sector. Journal of International Development 16: 63-80.

Ingenbleek, P. and A. Van Tilburg, 2009. Marketing for pro-poor development: deriving opportunities for development from the marketing literature. Review of Business and Economics 54: 327-344.

International Fund for Agricultural Development (IFAD), 2003. Promoting market access for the rural poor in order to achieve the Millennium Development Goals. International Fund for Agricultural Development, Rome, Italy.

International Fund for Agricultural Development (IFAD), 2004. Towards a global programme on market access initiative for mainstreaming innovation. Report of the pilot phase 6 October 2004. International Fund for Agricultural Development, Rome, Italy.

International Institute for Tropical Agriculture (IITA), 2001. Linking farmers to markets – overview from ACIAR. P.H. News No. 4 May, International Institute for Tropical Agriculture, Ibadan, Nigeria.

Kamara, A.B., B. Van Koppen and L. Magingxa, 2002. Economic viability of small scale irrigation systems in the context of state withdrawal: the Arabie scheme in the Northern Province of South Africa. Physics and Chemistry of the Earth 27: 815-823.

Kambewa, E.V., 2007. Contracting for sustainability: an analysis of the Lake Victoria-EU Nile perch chain. Wageningen Academic Publishers, Wageningen, the Netherlands.

Kherralah, M., C. Delgado, E. Gabre-Madhin, N. Minot and M. Johnson, 2002. Reforming agricultural markets in Africa. IFPRI Food Policy Statement No. 38. International Food Policy Research Institute, Washington, DC, USA.

Killick, T., J. Kydd and C. Poulton, 2000. Agricultural liberalisation, commercialisation and the market access problem in the rural poor and the wider economy: the problem of market access. Background paper for IFAD Rural Poverty 2000 Report, Rome, Italy.

Kirsten, J. and K. Sartorius, 2002. Linking agribusiness and small-scale farmers in developing countries: is there a new role for contract farming? Development Southern Africa 17: 503-529.

Leeflang, P.S.H. and W.F. Van Raaij, 1995. The changing consumer in the European Union: a 'meta-analysis'. International Journal of Research in Marketing 12: 373-387.

Madikizela, S.P. and J.A. Groenewald, 1998. Marketing preferences and behaviour of a group of small-scale irrigation vegetable farmers in Eastern Cape. Agrekon 37: 100-109.

Magingxa, L.L., 2006. Smallholder irrigators and the role of markets: a new institutional approach. PhD thesis, University of the Free State, Bloemfontein, South Africa.

Mansfield, B., 2003. Fish, factory trawlers, and imitation crab: the nature of quality in the seafood industry. Journal of Rural Studies 19: 9-21.

Mokitimi, N., 2007. Institutional constraints to horticulture production and marketing in Lesotho. Paper Presented at the Workshop for Inception of the Fund for Regional Innovative Collaborative Project (FIRCOP), Sept 2007, Johannesburg, South Africa.

Nel, E., T. Binns and D. Bek, 2007. 'Alternative foods' and community-based development: rooibos tea production in South Africa's West coast mountains. Applied Geography 27: 112-129.

Nel, E., T. Binns and N. Motteux, 2001. Community-based development, non-governmental organizations and social capital in post-apartheid South Africa. Geografiska Annaler, Series B: Human Geography 83: 3-13.

Oettle N., S. Fakir, W. Wentzel, S. Giddings and M. Whiteside, 1998. Encouraging sustainable smallholder agriculture in South Africa. Environment and Development Consultancy Ltd., Hillside, UK. Available at: http://www.eldis.org/vfile/upload/1/document/0708/DOC6491.pdf.

Omamo, S.W., 1998. Transport costs and smallholder cropping choices: an application to Siaya District, Kenya. American Journal of Agricultural Economics 80: 116-123.

Otsuki, T., J.S. Wilson and M. Sewadeh, 2001. Saving two in a billion: quantifying the trade effect of European food safety standards on African exports. Food Policy 26: 495-514.

Panisello, P.J. and P.C. Quantick, 2001. Technical barriers to hazard analysis critical control points (HACCP). Food Control 12: 165-173.

Ponte, S. and P. Gibbon, 2005. Quality standards, conventions and the governance of global value chains. Economy and Society 34: 1-31.

Ponte, S., 2005. Bans, tests and alchemy: food safety standards and the Ugandan fish export industry. DIIS Working paper no 2005/19. Danish Institute for International Studies, Copenhagen, Denmark. Available at: http://www.diis.dk/sw15843.asp.

Ponte, S., J. Raakjaer and L. Campling, 2007. Swimming upstream: market access for African fish exports in the context of WTO and EU negotiations and regulations. Development Policy Review 25: 113-138.

Powell, W.W., 1991. Neither market nor hierarchy: network forms of organization. In: G. Thompson, J. Frances, R. Levacic and J. Mitchell (eds.), Markets, hierarchies and networks: the coordination of social life. Sage Publications, London, UK, pp. 265-276.

Ruben, R., M. Van Boekel, A. Van Tilburg and J. Trienekens (eds.), 2007. Tropical food chains: governance regimes for quality management. Wageningen Academic Publishers. Wageningen, the Netherlands.

Statistics South Africa (SSA), 2001. Census reports. Government Printer, Pretoria, South Africa.

Stern, L.W., A.I. El-Ansary and A.T. Coughlan, 1996. Marketing channels, 5th ed. Prentice-Hall, Upper Saddle River, NJ, USA.

Thorbecke, E. and M. Nissanke, 2006. Introduction: the impact of globalization on the world's poor. World Development 34: 1333-1337.

Umhlaba Consulting Group, 2006. Feasibility study and sustainability plan for the emerging Alice-Kat citrus farmers. Amathole District Municipality, South Africa.

Urquhart, P., 1999. IPM and the citrus industry of South Africa. Gatekeeper Series No.86. International Institute for Environment and Development, Sustainable & Rural Livelihoods Programme, London, UK. Available at: http://pubs.iied.org/6338IIED.html.

Van der Meer, C.I.J., 2006. Exclusion of small-scale farmers from coordinated supply chains: market failure, policy failure or just economies of scale? In: R. Ruben, M. Slingerland, and H. Nijhoff (eds.) Agro-food chains and networks for development. Wageningen UR Frontis Series. Springer, Berlin, Germany.

Van Schalkwyk, H., J. Groenewald and A. Jooste, 2003. Agricultural marketing in South Africa. In: L. Nieuwoudt and J. Groenewald (eds.) The challenge of change: agriculture, land and the South African economy. University of Natal Press, Pietermaritzburg, South Africa.

Van Tilburg, A., J. Trienekens, R. Ruben and M. van Boekel, 2007. Governance for quality management in tropical food chains. Journal on Chain and Network Science 7: 1-9.

Verhallen, T., V. Wiegerinck, C. Gaaker, and T. Poiesz, 2004. Demand driven chains and networks. In: T. Camps, J. Diederen, G.J. Hofstede and B. Vos (eds.) The emerging world of chains and networks: bridging theory and practice. Reeds Business Information, Doetinchem, the Netherlands, pp. 129-146.

World Bank, 2000. World Development Report 2000/2001: attacking poverty. Oxford University Press, Oxford, UK.

World Bank, 2005. Food safety and agricultural health standards: challenges and opportunities for developing countries. World Bank Sector Report, World Bank, Washington, DC, USA.

Zeithaml, V.A., 1988. Consumer perceptions of price, quality and value: a means end model and synthesis of evidence. Journal of Marketing 52: 2-22.

11. Factors unlocking markets to smallholders: lessons, recommendations and stakeholders addressed

Aad van Tilburg and Ajuruchukwu Obi

11.1 Objectives and research questions of the book

The book aimed to present results of investigations conducted to ascertain to what extent South Africa's small-scale farmers can share in the expected gains of integration into the national and international markets. Related to that broad goal, it was also intended to identify what institutional and other reforms are necessary to enhance the effective and profitable participation of these farmers in the regional economy given the constraints that smallholder agriculture is facing in South Africa. These objectives are elaborated in Chapter 1. Following on that, the specific research objectives are partitioned into three levels, namely the micro, meso, and macro level. Micro-level research objectives seek to identify key production and marketing constraints faced by smallholders, and the investigation of the degree of participation of these smallholders in both input and output markets. The meso-level research objectives are to determine the kind of farmer-based structures and institutions needed to empower smallholder farmers to address their constraints, and to investigate the feasibility of governance systems that can be used in the supply chains of farm commodities produced by smallholders. The macro-level research objective is to recommend to stakeholders and policy makers on how to improve the institutional and policy environment of smallholders.

This chapter draws on the lessons learnt through the various chapters in the book, bringing the theory and learning together. An attempt is made to bring all the different strands together into the analytical and conceptual framework presented in Figure 11.1 which, as an adaptation of Figure 5.1 in Chapter 5, expands on similar concepts introduced and discussed in that chapter.

Figure 11.1 gives a framework catering for the needs of all possible role-players in the value chain. The figure is divided into three categories of farmers, namely subsistence, emerging commercial farmers, and commercial farmers. At each level the degree of involvement by government and/or private sector role-players would be different. The figure suggests that the main responsibility for support on the subsistence level should reside with public authorities. However, subsistence farmers have the potential to develop into emerging commercial producers as is depicted in the middle of Figure 11.1. Support to this group of farmers would be in the form of an alliance including government, private sector, academic institutions and commercial farmers' initiatives. Emerging farmers are not yet ready to enter the commercial market insofar as technology gathering and adoption, as well as management skills are concerned, yet they do not qualify as subsistence farmers any longer as targeted by international and governmental support programmes. As they move towards complete

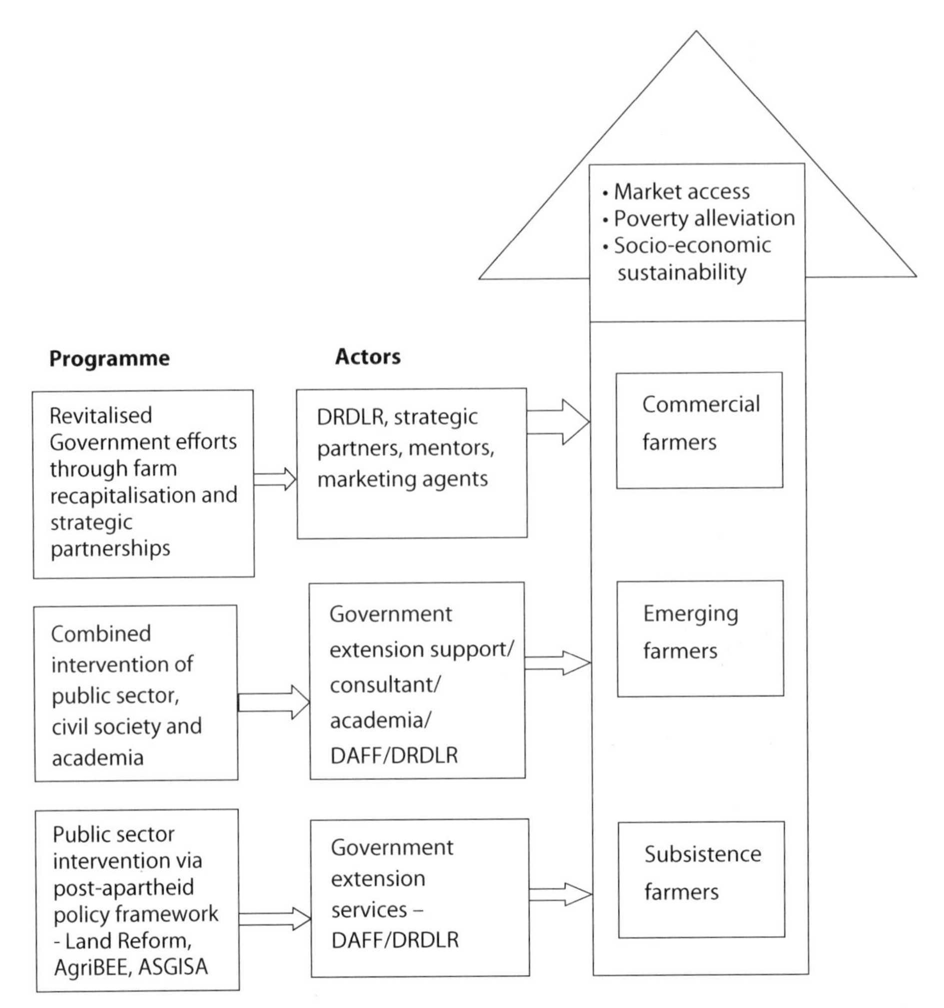

Figure 11.1. Developmental-transformational path under South Africa's post-apartheid agricultural restructuring programme (Adapted from Jordaan and Jooste, 2003).

commercialisation, the support functions performed by government could be transferred to other role-players. The key features in Figure 11.1 are used to summarise the analysis in each chapter of the book and to derive main lessons from the analysis (Box 11.1).

The conclusions of the various chapters are brought together into a concise synopsis (Table 11.1). Policy recommendations from each chapter are grouped for the relevant stakeholders: farmers, farmers' organisations, the agribusiness sector, NGOs and governmental bodies in Table 11.2. This will contribute to evidence-based decision making for planning and policy of responsive strategies for smallholders in South Africa and, probably, also in neighbouring countries. A summary of the analysis in each chapter is presented in Table 11.1.

Box 11.1. Key features used to summarise the analysis in each chapter of the book and to derive lessons learnt.

Category of farmers and stage of farmer development

Category:
- subsistence farmers;
- emerging commercial farmers;
- commercial farmers.

Stage of farmer development:
- S → E: from subsistence to emerging;
- E→ C: from emerging to commercial.

Variable(s) to be explained: Market access rate, or related variables.

Potential explanatory variables:
- Access to resources, e.g. human, social and economic capital of the farming household. Examples are bargaining power, conflict handling, participating in business networks, or the asset base of the household.
- Source(s) of institutional support originating, for example, from:
  - governmental institutions;
  - non-governmental institutions;
  - market institutions, e.g. related to market transparency;
  - financial institutions;
  - organised agriculture;
  - agribusiness;
  - co-operatives;
  - educational institutions;
  - physical or communication infrastructure.
- Type of governance in the value chain:
  - market-based;
  - contract-based;
  - trust-based;
  - mixed form.

Level of the analysis:
- Micro level.
- Meso level.
- Macro level.

Lessons learnt and policy recommendations

Table 11.1. Main elements of the analysis in each of the chapters of this book corresponding to the framework presented in Figure 11.1.

S→E; E→C[1]	Variable(s) to be explained	Explanatory variables[2]	Governance type of value chain (level of analysis)	Lessons and recommendations[3]
Chapter 2. Strategies for market access				
S→E	strategies for market access: high-volume and low value, or low volume and high-value	HSE capital, market transparency, group action and bargaining power, access to resources	market, contract, trust, mix (micro, meso)	form a multi-institutional task force to monitor the improvement of market access
Chapter 3. Factors for channel choice; Kat River area				
S→E	probability to participate in markets	HSE capital, institutional support, infrastructure	market (micro)	encourage collective action; contract farming; market transparency; value adding activities; investment in rural infrastructure
Chapter 4. Constraints to market access; Nkonkobe area				
S→E	probability to sell all produce offered	HSE capital, institutional support, infrastructure	market (micro)	encourage and actively facilitate access to market and price information, productive inputs and technical information
Chapter 5. Smallholders and livestock markets				
S→E	market access of livestock farmers	HSE capital, institutional support, infrastructure	market, contract (micro)	concerted action regarding: motivation; improved quality; market information; access to livestock identification facilities; marketing infrastructure; crime prevention; and appropriate institutional support.
Chapter 6. Contracting to unlock markets				
S→E	market participation rates	HSE capital, institutional support	contract (micro)	counteract monopsonistic situations; contract (new) niche products; Stimulate produce traceability for improved quality; facilitate access to credit and inputs; organise training and mentorship for emerging farmers

Table 11.1. Continued.

S→E; E→C[1]	Variable(s) to be explained	Explanatory variables[2]	Governance type of value chain (level of analysis)	Lessons and recommendations[3]
Chapter 7. Food retailing and agricultural development				
E →C	effects on: supplier and consumer prices; traditional retail outlets; regional economies (employment); the agrifood sector, e.g. access of smallholders to centralised procurement.	impact of retail sector development on consumers and agriculture	contract (meso)	positive effects: smallholder access to good quality produce; convenience shopping and lower prices for basic household goods negative effects: exclusion of smallholders and traditional markets by specialised retailers
Chapter 8. Unlocking credit markets				
S→E	access to credit markets	demand-led versus supply-led approach, sustainability, innovation, type of collateral	contract, trust (meso)	form a network of MFI's; provide credit conditional to the sales of produce; involve the Land Bank in supplying credit to MFI's; financial services need to include credit, saving, insurance and money-transfer; offer decentralised financial services
Chapter 9. Governance structures for supply chain management				
S→E	governance structure	constraints, conflicts, transaction costs	market, contract, trust, mix (micro, meso)	take into account: needs of farmers and other chain participants; sources and destinations of principal foods in the district; arrangements under which transactions take place; the elements of producer-retailer contracts; responsibility for municipal regulatory mechanisms; types of actors having access to the fresh produce market; links between major retail chains and Fresh Foods Markets; the influence of bodies enforcing Good Agricultural Practices

Table 11.1. Continued.

S→E; E→C[1]	Variable(s) to be explained	Explanatory variables[2]	Governance type of value chain (level of analysis)	Lessons and recommendations[3]
Chapter 10. Market access and governance				
S→E	access to assets, knowledge, services and markets	governance mode	market, contract, trust, mix (micro, meso)	promote training of smallholders; encourage farmer organisations to secure credit; improve collective action for bargaining power and scale economies; distribute non-productive land to those interested; create a land market reflecting the economic value of land; rebuild smallholder support services; facilitate access to market information; introduce incentives for transport contractors; improve the road infrastructure; initiate a multi-stakeholder platform

[1] Farmer development: S = subsistence farmers; E = emerging commercial farmers; C = commercial farmers.
[2] HSE capital = human, social and economic capital.
[3] MFI = microfinance institution.

11.2 Results of the analysis in the chapters of the book

The content of Table 11.1 is briefly discussed in this section. In Chapter 2 dealing with strategies to improve smallholders' market access, insights were obtained from sources in economics, management and marketing and several case studies on problems that smallholder farmers in South Africa face when trying to improve their market access. Based on these insights, strategies were developed to deliver market access to smallholders, but considerable effort is needed to scale up successful examples. It is advised that smallholders and their stakeholders in value chains form a multi-institutional task force to reach this goal.

In Chapter 3, dealing with institutional and technical factors influencing marketing channel choices amongst smallholder and emerging farmers in the Kat river valley, smallholder farmers do not use output markets effectively due to technical and institutional factors.

Their choices are either to market their produce on formal or informal markets or not to participate in markets. Results show that market information, expertise in grades and standards, contractual agreements, social capital, market infrastructure, group participation and tradition significantly influence household marketing behaviour. Lessons learnt are that smallholders tend to be trapped and continue to operate within given market constraints and, consequently, do not receive rewarding incomes from their agricultural activities. Recommendations encourage collective action, contract farming, the availability of market information, value adding activities, investments in rural infrastructure and government support policies in rural areas. These strategies are now widely recognised to be pre-requisites for effective market participation for resource-poor farmers especially where the emphasis is on relieving capital constraints through group action.

In Chapter 4, dealing with technical constraints to market access for crop and livestock farmers in Nkonkobe Municipality, Eastern Cape Province, market access is dependent both on production and marketing issues. Access to market and price information, and productive inputs, are crucial to smallholder development. It is recommended that smallholder farmers be furnished with technical information that they can conveniently process and use. One way to achieve this could be to reinstate the support services as provided in the past by the public sector. Recent public sector pronouncements highlight the urgency to minimise, if not completely eliminate, farm failures and a farm recapitalisation programme linked to strategic partnerships is being promoted by the newly created Department for Rural Development and Land Reform. It is probably too early to assess the impact of that programme.

Chapter 5 is dealing with smallholder and livestock markets mainly in the former 'reserves' or 'homelands' where land ownership tends to be communal, and in which overgrazing and deterioration of the vegetation has become a serious problem. Traditional farmers have a low take-off in terms of livestock sold due to the restrictions to market access, very small herds, absence of savings institutions and investment opportunities which caused livestock to be the only way of accumulating capital and to be the only source of readily transportable wealth. Poor marketing infrastructure and cultural reasons have together with the absence of savings facilities contributed to overstocking, particularly during droughts. After analysing case studies, the conclusion is that smallholders' access to livestock markets can be improved by concerted attention to, and action regarding: motivation; improved quality and condition of livestock; market information; access to livestock identification facilities; marketing infrastructure; crime prevention; and appropriate institutional support.

Chapter 6 discusses the potential of contracting to unlock markets to smallholder farmers. Smallholders have historically shown low rates of participation in the marketing of agricultural products. This problem has to be overcome in a period of rapid change in the retailing of food products which presents important challenges, some of which have been caused by low skill levels among, and high transaction costs to, small producers. Contracting provides the smallholder with a definite outlet for a specified product, while the processor/

marketer gains the advantage of an assured supply of the product desired at a previously arranged price, thus reducing risk to both parties. However, the success of contracting can be hampered by problems of opportunistic behaviour, relationship problems between supplier communities and buyers, relationship problems within supplier communities, moral hazard and free rider problems. Asymmetric power relationships with the buyer in a monopsonistic position often constitute another hazard. These problems can be overcome by good advance planning and continual attention to developments as is borne out by examples of communities engaged in contract farming.

Chapter 7 is dealing with food retailing and agricultural development in South Africa over the past few decades and the involvement of the global food retail sector in a dynamic and competitive industry. South Africa's food retail industry is defined by an intense level of competition, concentration, and a unique both South African and African customer base, which vary from the poor in rural areas to the rich in cities. The food retail industry in South Africa is dominated by large supermarket chains alongside smaller chains, independent supermarkets and an informal market. It is concluded that the evolving retail industry can have either positive or negative effects on the regional economy. It could positively influence the area by providing local people with access to good quality produce, convenience shopping and lower prices for basic household goods. The negative aspects include the exclusion of traditional markets for fresh produce by specialised retailers, reduction of market access possibilities for local and small-scale producers and suppliers, and thus limiting the potential for money generation within the region. Farmers providing quality products on a continuous basis at the required volumes are gaining significantly with a secure market and prices.

Small farmers and small entrepreneurs have to cope with serious problems regarding access to finance (Chapter 8). A lack of collateral inhibits banks and other formal financial institutions to supply credit to these entrepreneurs, and the little credit that is available comes at a high cost. In the supply-led approach to credit, governments and donor agencies invested money on mostly government-run institutions to supply cheap credit. This approach led to high rates of delinquency and was characterised with very poor loan recovery. It also discouraged savings. The demand-led approach, driven by savings mobilisation, financial sustainability and institutional innovation has less government and donor involvement. It practices group lending, a cornerstone of microcredit in many countries. Many types of microcredit institutions exist with rotating savings and credit associations as the predominant forms of informal financial institutions. These consist of groups of individuals who form associations to save money, share risks and borrow money. Group accountability is taken as collateral.

Chapter 9, dealing with governance structures for supply chain management in smallholder farming systems, discusses technical and institutional constraints that confront smallholders and impede their market access and frequently manifest in conflict situations. The agricultural market is composed of a large number of participants, both sellers and buyers, whose goals are in conflict by definition – one well-known case being their differential attitudes to high food prices. Additionally, there are multiple other interests in the supply chain, e.g. those of mediating entities, which must be taken into account. Optimal supply chain governance has as its principal outcomes the enhancement of market access for smallholders, satisfaction of consumer expectations through better prices and quality, and overall welfare improvements for the community. Many farming and marketing systems have in-built structures and institutions such as contracts and laws, business organisations, and cooperative enterprises, as well as numerous arrangements, that act to varying degrees to perform these functions, although in many situations these systems and procedures are poorly developed. The chapter presents insights gained in the key determinants of the commodity mix in the farming system, the range of participants in the supply chain, the existing regulatory mechanisms, and the nature of transactions that move foods along the supply chain. The implications of these arrangements for market access, poverty alleviation and socio-economic sustainability are evaluated to contribute to policy formulation for smallholder development through enhanced market access.

Chapter 10 explores the relationship between smallholder access to markets and the governance mode in supply chains by examining how the governance mode influences access to assets, market information, services and remunerative markets. This is done by reviewing the literature and by analysing two typical case studies: one case study looks at domestic distribution in a case of smallholder access in irrigation schemes in South Africa, the other case is on market access of small scale fishers to an international supply chain. The analytical framework presented helps to derive several lessons from these case studies.

11.3 Recommendations in relation to stakeholders addressed

As has been clearly articulated throughout the book, achieving enhanced market access for small-scale farmers was identified as a crucial need which would obviously involve special arrangements to actualise. The theoretical literature foresees immense gains from integration into national and international markets and there are time-tested opinions on what institutional and other reforms are necessary to enhance the profitability of those participating in the regional economy against the backdrop of the overwhelming and well-documented constraints that smallholder agriculture currently faces in South Africa. The book points to the factors that influence the process of inclusion or exclusion of smallholder farmers in national or international markets and it addresses stakeholder groups that are considered or expected to be able to facilitate effective and profitable participation in the national or regional economy.

Table 11.2 builds on the recommendations in the various chapters of the book in relation to the stakeholders addressed. The large number of recommendations, thirteen on the whole, reflect the enormity of the issues identified. It is however feasible to aggregate them according to the stakeholder group targeted by a specific recommendation. From Table 11.2, it is clear that initiatives on improvements which benefit groups of smallholder farmers

Table 11.2. Recommendations in relation to stakeholders addressed.

Recommendations	Chapter(s)	Stakeholders addressed
Encourage need-driven empowerment and support of smallholders	5, 6, 9,10	Farm schools Farmers organisations of smallholders Irrigation authorities Extension services Agribusiness sector NGO's
Encourage access to land, farm inputs, technical information, microfinance, means of transport	4, 6, 8, 10	Land markets Land Bank Supply chain members Networks of microfinance Institutions
Encourage collective action of smallholders	3, 10	Smallholders Farmers organisations Agribusiness sector NGO's
Encourage value adding activities	3	Farmers organisations of smallholders Extension services Agribusiness sector NGO's
Stimulate a safe operational environment	5, 9	Sector organisations Public authorities
Encourage market transparency	3, 4, 5, 10	Commodity exchanges Agribusiness sector Public authorities
Counteract monopsonistic situations	3, 6, 10	Farmers organisations of smallholders Public authority
Encourage vertical coordination in the supply chain	3, 6, 9	Farmers organisations of smallholders Supply chain members NGO's
Improve good agricultural practices and product quality	5, 6, 7, 9	Retail sector Supply chain members NGO's
Improve product traceability	5, 6	Retail sector Supply chain members
Be pro-active with respect to developments in the retail sector regarding inclusion and exclusion of smallholders	7, 9	Farmers organisations of smallholders Agribusiness sector Retail sector

Table 11.2. Continued.

Recommendations	Chapter(s)	Stakeholders addressed
Encourage investments in the rural infrastructure including markets for land, inputs, finance and outputs	3, 5, 8, 9, 10	Public authorities Agribusiness sector Finance sector
Form a multi-institutional task force to improve market access in each agribusiness sector	2, 10	Farmers organisations of smallholders Agribusiness sector Research institutes Government NGO's

require usually a multi-institutional effort, for example, coordinated by a task force or similar collective action arrangements. Several examples of this approach have been documented in this book.

To encourage both need-driven empowerment and support of smallholders, farm schools, irrigation authorities, extension services and the agribusiness sector can offer vocational training, notably in the lean season of the farmer's labour calendar. Representative bodies, such as farmer organisations and NGO's, might jointly coordinate and monitor such training programmes to obtain the maximum benefit out of an effort. It is noted that government is actively promoting youth skills development across various disciplinary areas through aggressive internship programmes for fresh university graduates and those from the polytechnics and technical colleges. Conscious effort to align the curricula for such training and internships schemes to real needs on the ground will contribute significantly towards bridging the skills gaps in the country and go a long way towards strengthening municipal level implementation capacities that have come under strong criticisms in recent years. The spate of service delivery protests that the country has witnessed lately can be linked to this acute skills gap.

To encourage access to land, farm inputs, technical information, microfinance and means of transport, first and for all (land, output, inputs, credit) markets need to operate properly. The Land Bank, supply chain members and networks of microfinance institutions are expected to initiate, coordinate and monitor proper activities preferably in joint or bilateral actions. Expertise in correctly assessing these needs is also crucial and these organisations have a role to play in that regard.

To encourage collective action of smallholders, initiatives need to be taken by groups of smallholders, their representative bodies in the form of farmers organisations or NGO's, and the agribusiness sector as partner in a well-coordinated value chain. Where prior experience is either lacking or limited, a good deal of mentorship is required to kick-start these initiatives and monitor their operation as closely as possible.

Value adding activities can be encouraged by inventive smallholders, their representative organisations, extension services, preferably in joint and coordinated actions. This is what the strategic partnership arrangements should do, among their other activities in farm recapitalisation and revising failed and failing farms.

To improve the safety of the operational environment of smallholders, sector organisations can intermediate to obtain adapted insurance services on health, disasters including theft and fires; and rural communities and public authorities need to cooperate to reduce theft of property, e.g. cattle or crops.

To encourage market transparency, freely accessible markets such as auctions, warehouses for crop storage – to obtain collateral for credit – and commodity exchanges need to be established or improved in a joint effort of farmers organisations, the agribusiness sector and public authorities.

To counteract monopsonistic situations, farmers or their organisations need to improve their countervailing power by joint action and public authorities need to improve and enforce antitrust regulations.

To encourage vertical coordination in the supply chain, both initiatives and joint action are required of farmers or their representative bodies and supply chain members. To improve good agricultural practices and product quality, a joint effort of all supply chain members from farmer to retailer is required to establish standards, certification procedures and vocational training for farmer groups. Product traceability, supported by all supply chain members, is a good way to be able to monitor agreed quality standards.

Farmers or their representative bodies need to be aware and pro-active with respect to changes in consumer behaviour and the response of retail chains to these changes. Pro-activity can be more easily realised in well-coordinated supply chains than in other supply chains.

Well-coordinated investments in rural infrastructure can improve the performance of the smallholder sector substantially. Initiatives should be taken by the agribusiness and finance sector in close consultation and cooperation with public authorities.

An overall recommendation is that individual stakeholder groups might be able to improve market access of smallholders, but that joint actions of stakeholder groups such as farmer organisations, the agribusiness sector, research institutes, NGO's and public authorities can and should be more effective and efficient in reaching stated goals.

About the authors

Alemu Zerihun Gudeta holds a PhD in Agricultural Economics from the University of Free State, South Africa. He is currently a policy analyst at the Development Bank of Southern Africa (DBSA), a position he has been holding since July 2007. He is also affiliate associate professor at the University of the Free State, South Africa. Prior to that he lectured at the University of the Free State (2002 to 2007) and at the Haramaya University, Ethiopia (1992 to 2001). Dr. Alemu's expertise includes agricultural policy analysis, agricultural marketing, applied time series econometrics, and international trade. He has widely published in local and international peer reviewed academic journals and has delivered papers at a number of national and international conferences.
E-mail: zerihuna@dbsa.org

Gavin Fraser is Professor of Economics at Rhodes University in Grahamstown. Previously he was Professor of Agricultural Economics at the University of Fort Hare. He has a Master of Commerce degree from Rhodes University and a PhD (Agric) degree from the University of Stellenbosch, both in Agricultural Economics. His research interests are in the fields of agricultural and rural development, institutional economics and environmental and resource economics. He published in Agrekon, Development Southern Africa, Review of Southern African Studies, South African Geographical Journal, Journal of Agricultural and Food Economics and African Journal of Agricultural Research. In addition, he has produced a number of research reports and book chapters.
E-mail: g.fraser@ru.ac.za

Jan A. Groenewald has been extraordinary professor of Agricultural Economics at the University of the Free State since 1998, having retired in 1997 as professor at the University of Pretoria. His earlier research was mainly in the fields of farm management, production economics and agricultural policy, but in later years, the emphasis shifted from farm management and production economics to agricultural development and marketing. He has published approximately 200 articles in journals such as Agrekon, South African Journal of Economics, South African Journal of Economic and Management Sciences, South African Journal of Agricultural Extension, Development Southern Africa, Agricultural Systems and Journal of Agricultural Economics. He also contributed chapters to some books and co-edited four books.
E-mail: groenewald@netralink.com

Bridget Jari is a PhD student in the Department of Economics and Economic History, Rhodes University. She completed her BSc Agriculture (Agricultural Economics) *cum laude* and MSc Agriculture (Agricultural Economics) degrees at the University of Fort Hare in 2006 and 2008, respectively. Her MSc thesis investigated the institutional and technical factors that influence marketing among emerging and small-scale farmers. The research was carried out in the Kat River Valley, Eastern Cape Province of South Africa, and the main findings were based on information provided by local-level emerging and small-scale farmers. Part of the MSc thesis was published in the African Journal of Agricultural Research in 2009. Her PhD research focuses on the impact of Fairtrade on agricultural producers in South Africa. Bridget's interests are in development economics, especially on issues related to agriculture, markets and trade.
E-mail: beejarry@yahoo.com

André Jooste is currently senior manager of the Market and Economic Research Centre at the National Agricultural Marketing Council in South Africa and an Affiliate Professor at the University of the Free State. His research interest includes conducting market, policy, rural and industry analysis, as well as international trade issues and their implications from a country and firm point of view. He published in Agrekon, South African Journal of Economic and Management Sciences, South African Journal of Agricultural Extension, Journal of Development Perspectives, Journal of Policy Modeling and Quarterly Journal of International Agriculture. He contributed to chapters in several books and technical reports, amongst others, Agricultural Policy Reform in South Africa (1998, Francolin Publishers (Pty) Ltd), Agricultural Marketing in the New Millennium (2003, First National Bank), and The Challenge of Change: Agriculture, Land and the South African Economy (2003, University of Natal Press).
E-mail: andre@namc.co.za

Andries Jordaan holds degrees in BSc Agric and MSc Agric (Agricultural Economics) from the University of the Free State and a Honours degree in Agricultural Extension from the University of Pretoria. He is a registered professional natural scientist and currently a PhD fellow at the University of the Free State. He has published several scientific and popular articles and presented papers at various national and international Conferences. He was project leader for several national and international developmental projects. He traveled extensively in Africa as an advisor on agricultural development and disaster risk reduction issues. Mr. Jordaan is the Director of the Disaster Management Training and Education Centre for Africa (DiMTEC) at the University of the Free State where he also lectures in disaster risk reduction, development and environmental economics.
E-mail: jordaana@ufs.ac.za

Emma Verah Kambewa got her PhD from Wageningen University in 2007 on research in the value-added chain of fish from lake Victoria to Europe. She has recently been a social scientist with the WorldFish Center, Malawi. Currently she is working with AGRA (Alliance for a Green Revolution in Africa) based in Kenya. Her work is mainly in Southern Africa, including Malawi, Mozambique, Tanzania and Zambia. Her research interests include channel governance including linking smallholder producers to markets, sustainability of natural resources and corporate social responsibility in marketing channels and value chains. She has published in the Journal of Macro-Marketing and contributed several book chapters on Corporate Social Responsibility or governance of value chains.
E-mail: vkambewa@hotmail.com

Jacobus Klopper obtained his Master's degree in Agricultural Economics at the University of the Free State, Bloemfontein, South Africa in 2009. During his studies he was a research assistant to Dr. Kobus Laubscher in the Department of Agricultural Economics, UFS and also of the Dean of the Faculty of Natural and Agricultural Sciences, Professor Herman van Schalkwyk. He assisted in scientific research projects and the writing of both scientific and research papers. His MSc thesis titled 'Mainstreaming of smallholder irrigators: The case of Taung irrigation scheme, North West, South Africa' focuses on factors influencing the success potential of small-scale farmers. The study focuses on the role of private sector involvement in helping small-scale farmers to improve their success potential and thus mainstreaming them into the South African economy. After completing his studies he joined a company involved in trading of agricultural commodities by means of futures and options on the South African commodity market SAFEX as well as the Chicago Board of Trade CBOT. He was also involved in export of agricultural commodities to neighbouring African countries. He is currently the Regional Manager of Agriculture for the Free State and Northern Cape provinces for Nedbank.
E-mail: kootk@nedbank.co.za; klopperkoot@hotmail.com

Litha Magingxa is Lead Researcher for the Land and Agricultural Development Bank of South Africa (Land Bank). His research interests include market access for small-scale farmers, agriculture and rural development as well as natural resource policy developments including land reform. His work on market access was the subject of his PhD thesis at the University of the Free State. He has written several journal publications. He has also contributed to several research reports and presented his work in national and international conferences.
E-mail: llmagingxa@landbank.co.za

Ajuruchukwu Obi is an Associate Professor in the Department of Agricultural Economics and Extension of the University of Fort Hare in Alice, Eastern Cape Province of South Africa. Professor Obi obtained his degrees in Agricultural Economics and Economic Policy Management from the University of Nigeria, University of the Free State (South Africa) and McGill University (Canada). He has previously worked for the University of Nigeria, International Labour Organisation (ILO), the United Nations Development Programme (UNDP), and the United Nations Volunteers (UNV), among other institutions. His main research interests include institutions, agricultural land prices and land reforms, crop-livestock integration in smallholder agriculture, water use in agriculture, value chain analysis, technology adoption and technical and institutional constraints to smallholder farming. He is publishing in the fields of agro-food chains, constraints to smallholder farming, and crop-livestock integration.
E-mail: aobi@ufh.ac.za

Peter Paul Takawira Pote is a Zimbabwean national pursuing his PhD in Agricultural Economics at the University of Fort Hare, having obtained an BSc and MSc in the same discipline from the same University. He is currently conducting research into the factors determining the integration of small-scale farmers into the mainstream agri-food systems of South Africa. Mr Pote has presented his research findings in many national and international conferences. He teaches Seminar Writing and Marketing of Agricultural Products at the undergraduate level at the Department of Agricultural Economics and Extension of the University of Fort Hare, Alice.
E-mail: potepeter@gmail.com

Lindie Stroebel (previously Botha) is the manager of economic intelligence at the Agricultural Business Chamber based in Pretoria, South Africa. The Agricultural Business Chamber is a member organisation for businesses in the South African agricultural sector. Lindie specialises in the field of competitive positioning of agribusinesses and acts on various councils, forums, committees and task groups to negotiate for and establish an enabling policy environment for agribusinesses. In 2006 she obtained her MSc Agric degree with distinction at the University of the Free State, where she also worked as a researcher in agricultural economics for Prof. Herman van Schalkwyk. She also lectured in agricultural development and policy.
E-mail: lindie@agbiz.co.za

Aad van Tilburg retired in 2010 from Wageningen University as associate professor in marketing. His research interests included the functioning and performance of market actors, markets, marketing channels and value chains. He published in Agribusiness, Agricultural Economics, European Review of Agricultural Economics, Journal of Business Venturing, Journal of Development Economics, Journal of Regional Science, Journal of African Economies, Journal on Chain and Network Science and Netherlands Journal of Agricultural Science. He was co-editor of several books including Agricultural Marketing and Consumer Behavior in a Changing World (1997, Kluwer Academic Publishers), Agricultural Marketing in Tropical Africa (1999, Ashgate Publishing), Agricultural Markets beyond Liberalization (2000, Kluwer Academic Publishers) and Tropical Food Chains (2007, Wageningen Academic Publishers).
E-mail: aad.vantilburg@wur.nl

Herman D. van Schalkwyk is Rector of the Potchefstroom Campus of the North-West University in South Africa since February 2010. Prior to this appointment he was Dean of the Faculty of Natural and Agricultural Sciences at the University of the Free State (since 2003). He obtained his PhD degree in Agricultural Economics at the University of Pretoria in 1995. He started lecturing at this university in 1993 after working as agricultural economist at the Standard Bank of South Africa for two years. He became associate professor at the University of the Free State in 1996 and full professor and head of the Department of Agricultural Economics in 1998. Prof. Van Schalkwyk published more than 70 articles in various scientific journals and about 117 research and project reports. He delivered more than 75 papers at local and international conferences. He advises many provincial and national institutions also on agricultural development issues. He also serves as deputy chairperson of the Board of the Land Bank and often acts as consultant for the World Bank and other South African institutions.
E-mail: herman.vanschalkwyk@nwu.ac.za

Keyword index

Printed in the United States
by Baker & Taylor Publisher Services